Ham... ...als

Hammond's Farm Animals

Fifth edition

Revised by

J Hammond Jr
Former School of Agriculture, University of Cambridge

J C Bowman
Natural Environment Research Council, Swindon

T J Robinson
Department of Animal Husbandry, University of Sydney

Edward Arnold

© J. HAMMOND JR., J. C. BOWMAN, T. J. ROBINSON, 1983

First published in 1940
by Edward Arnold (Publishers) Ltd
41 Bedford Square, London WC1B 3DQ
Reprinted 1941, 1943, 1944, 1946
Second edition, 1952
Reprinted 1955
Third edition, 1960
Reprinted 1962, 1965
Fourth edition, 1971
Reprinted 1976, 1978
Fifth edition, 1983

British Library Cataloguing in Publication Data
Hammond, *Sir* John
  Hammond's farm animals.—5th ed.
  1. Stock and stock breeding
  I. Title   II. Hammond, John, *1917–*
  III. Bowman, J. C.   IV. Robinson, T. J.
  636.08′2   SF61

ISBN 0 7131 2848 8

Printed in Great Britain by Butler & Tanner Ltd, Frome and London

Contents

Preface to the Fifth Edition

The decade that has elapsed since the appearance of the fourth edition of Hammond's *Farm Animals* has witnessed major advances in practical animal husbandry. Techniques which were at an experimental stage 10 years ago are now in commercial use, some on quite a large scale. Also with the development of more sophisticated forms of animal production in the 'third world', there is increasing interest in a number of types of livestock such as the buffalo, which hitherto have been relatively neglected, and there is increased interest in the goat, for fibre, milk and meat.

This new edition, which like the last is a tribute to John Hammond, attempts to update the fourth edition and, in addition, includes brief accounts of some characteristics of buffalo and goats. While the previous revision of Hammond's classic was a challenge, this one has been even more so, and the authors are acutely aware of their limitations in covering, with simplicity and authority, the area involved.

The first edition of *Farm Animals*, published in 1940, was based upon two series of lectures given by Hammond 'to inform the student and animal breeder of some of the results of recent scientific research and its application to the practical problems of livestock production.' Subsequent revisions were made by Hammond in 1952 and 1960 and another, in which two of the present authors (J. Hammond, Jr. and T. J. Robinson) were involved, appeared in 1971. In the current revision, as in the last, there has been no necessity to change the original format, and it is quite extraordinary to be made aware—once again—of how far ahead of its time was that slim volume of 1940. Some Hammond concepts have been challenged, and dissection data based on anatomical rather than arbitrary jointing techniques have necessitated a revision of a few of his basic ideas on growth and development of meat qualities. But they have not substantially altered the practical applications.

Hammond was a pioneer in modern animal science and he was essentially a practical man. It is rewarding therefore to be able to record in this book the many advances which have been made in areas with which he was associated. Control of breeding, artificial insemination with frozen semen, embryo transfer, early pregnancy diagnosis, artificial induction of lactation, feeding for optimum meat characters, objective carcase appraisal, performance recording and progeny testing are but a few of the areas of his personal interest and involvement which are now part of the weft and warp of modern animal industry.

As in earlier editions, many references have been given to the literature. Books and reviews are listed separately at the end of each chapter. Unfortunately it has been necessary to omit some references cited in earlier editions but important original sources have been retained. Regrettably, much that is published in some areas is largely repetitive and today's student needs to be reminded that modern animal production is based on principles many of which were elucidated decades ago.

Responsibility for revision of Part 1 was undertaken by T. J. Robinson and J. Hammond, Jr., and for Part 2 by J. C. Bowman. Our thanks are due to all those authors whose work has been cited in the text and illustrations. Much of the early basic work cited in this book was carried out by Hammond's students and associates at the Animal Research Station, Huntingdon Road, Cambridge and it is appropriate to acknowledge the assistance given them by Messrs C. Williamson, J. Pike, A. F. Smith (all now deceased) and G. Pluck. We are grateful to the publishers for their assistance in many ways.

1983   J.C.B.   J.H.Jr.   T.J.R.

Sir John Hammond, *C.B.E., F.R.S.*

JOHN HAMMOND (1889–1964), animal scientist, was born 23rd February 1889, at Briston, Norfolk, on the farm of his father, Burrell Hammond, a tenant of Lord Hastings. He was the eldest of four children and was christened John after his grandfather who, besides being a farmer and a veterinarian, was one of the founders of the Red Poll breed of cattle. His mother was Janette Aldis, daughter of a schoolmaster with his own school in the county at East Dereham.

Hammond received his schooling at Gresham's School, Holt, and at

Edward VI (Middle) School, Norwich. His Latin having failed to gain him admission to the Royal Veterinary College, Hammond was sent to Cambridge, on the advice of T. B. Wood (q.v.) in 1907 to study Agriculture. After taking the Natural Science Tripos, recommended by his tutor, he obtained the Diploma in Agriculture, with distinction in all biological subjects. On top of his farming background, his university training in pure and applied biology was to enable him in his life's work to change the traditional study of Animal Husbandry into an empirical science of Animal Production, of which he was the creator; and his coming under the influence of one of his teachers, F. H. A. Marshall (q.v.) was to give direction to his research, particularly in the physiology of reproduction.

At the outbreak of war in 1914 Hammond joined the 7th Battalion of the Royal Norfolk Regiment and served as Captain and Company Commander with the B.E.F. in France until invalided home in 1916. Later he was Staff Captain, 201 Infantry Division.

Hammond's career in animal science began in earnest after the war with his appointment as physiologist in the Animal Nutrition Institute. He later became Director of the Animal Research Station, which he remained until retiring in 1954. His budget for research was meagre, never to exceed £10 000 p.a., and so he tried out his ideas on strains of rabbits inbred for characteristics he wished to study in farm animals. Even for the latter – for example, horses and cattle – he chose diminutive breeds, such as Shetland pony mares and Dexter cattle, to increase the numbers he could afford to keep. With these and collections of pigs and sheep, his researches encompassed the animal life cycle from fertility and conception through pregnancy to birth, growth and development, lactation and the inheritance of the next generation. His ideas and methods are well illustrated in his books, *The Physiology of Reproduction in the Cow* (1927) and *Growth and the Development of Mutton Qualities in the Sheep* (1932).

True to his dictum that 'Science isn't Science until it is applied', Hammond promoted the results of his research in many ways. From those on growth in meat animals he worked with producer organizations at home and abroad in developing standards for carcase assessment which are now in international use. His early interest in artificial insemination, officially discouraged for a time, led to the formation of the country's first Cattle Breeding A.I. Centre at Cambridge in 1942, which became the model for a new breeding system, now developed nationally, for the genetic improvement of cattle for milk and beef.

Hammond, although very tenacious of his ideas for livestock improvement, was by nature extremely gentle and kind, and incapable of making an enemy. With his very tall, spare build and his healthy complexion of a countryman, he was an outstanding figure among gatherings of farmers and scientists among whom he was equally at home. He was an inveterate world traveller and invitations from many Governments and national organizations of producers to survey their animal industries provided him with opportunities which he relished. Whenever possible he went by train and in daylight, breaking his journey each evening, to enable him to make diary entries of the farming along his route. Hammond rowed for his

College (Downing), keeping an interest in the Boat Club throughout his life, and remained active and fit by cycling daily to and from his work and by gardening intensively to produce fruits and vegetables from which his family and the families of his research students benefited in abundance. His interest in genetics overspilled from the laboratory to the garden, where he bred numerous strains of polyanthus.

In 1916 Hammond married Frances Mercy, daughter of John Goulder, farmer, by whom he had three sons. He was appointed C.B.E. (1949) and received his Knighthood (Kt. Bachelor) in 1960. He was elected a Fellow of the Royal Society in 1933 and was made a Fellow of Downing College in 1936. Honorary Doctorates were conferred on him by the Universities of Iowa (1932), Louvain (1953), Durham (1956), Copenhagen (1958), Leeds (1961), Krakow (1963) and by the Hochschule für Bodenkultur, Vienna (1952). He was made a Commander of the Order of Orange-Nassau (1946) and Commanda al Merito della Republica Italiana (1954) and was elected Foreign Member of many Academies of Agriculture and Veterinary Science.

[Sir Wm. Slater & Dr Joseph Edwards, *Biographical Memoirs of Fellows of the Royal Society,* Vol. II, November 1965: Elizabeth O. Cockburn, *Journal of Animal Production*, Vol. 4, Part I, 1962: personal knowledge]

J. Edwards

Part 1
Fertility and Growth

1 General principles—reproduction

An animal is the product of the interaction between its heredity and its environment. The hereditary make-up is determined at the moment of fertilization of the egg by a sperm, by the combination of genetic material from each parent. This genetic material then becomes active in the various cell types as the individual unfolds, and the interaction between the different cells, and with the external environment, results in the development of the mature animal.

Perhaps the most obvious instance of the moulding effect of environment is that of childhood experience on adult personality; but of course physical characters are affected also, though not all equally readily. Their 'heritability' (Chapter 10) is a measure of the ease with which genetic influences are modified. The latter part of this book deals primarily with genetic aspects of animal improvement, and the earlier section mainly with the applications of physiology. The aim of these first two chapters is to outline the general principles of reproduction and growth, and to indicate possible applications; though what one imagines as a practical possibility depends partly on knowledge of the range of problems in livestock production, partly on vision of how basic knowledge could be applied, and partly on ignorance of gaps in basic knowledge (which may only be revealed when the attempt is made to apply what is known).

Transmitter substances and hormones

As a fertilized egg divides, different parts of its genetic make-up are activated in what become different cell types. Different cell types interact to form organs, and the activity of these organs is coordinated in two ways—through messages either conveyed along nerve fibres which end close to the cells they regulate, or carried to them in the blood and tissue fluids from elsewhere in the body.

In general, the message passes by the release of a special transmitter substance which affects other cells only if they have on their surface receptors with an affinity for that particular transmitter. Combination of the receptor with the transmitter substance then either stimulates a function already possessed by that cell, or activates genes hitherto dormant—effectively inducing a new cell type.

Hormones are transmitter substances with targets at a distance from their site of release; necessarily they are discharged in sufficient amounts, and are sufficiently stable, to sustain effective concentrations after transport through the whole course of the blood circulation. Otherwise the transmitter could have only a localized action.

Between the purely locally effective substances (such as that of a nerve controlling a muscle) and the hormone, there is an intermediate type of transmitter—one which is carried in the blood, but not diluted in its whole volume, because the path of the circulation from its site of origin leads directly to the organ affected.

Through these transmitter-receptor mechanisms one may modify normal functions in several ways. Hormones (extracted from the endocrine glands that form them, or synthesized if their structure is not too complex) may be administered, or drugs may be given which prevent the natural formation of particular transmitters. Compounds may be synthesized with structures analogous to that of a natural hormone. Such analogues can have a variety of effects. They may be more potent than the natural hormone, either because they are less readily destroyed, or because they bind more strongly to the receptor; or they may block the action of the natural hormone by binding to the receptor without inducing the effect which normally follows.

Most hormones are inactive when given by mouth, being destroyed by the digestive enzymes or by bacteria in the gut, but some analogues may be absorbed unchanged. On the other hand, injected substances may have an effect through the activity of a metabolite, the injected substance having no activity, or even a different one.

The process of reproduction

The male sex cells, the spermatozoa, are produced in the tubules of the testis and pass to a long coiled duct, the epididymis, in which they mature and are stored until required. The sperm are not fertile as they leave the testis, but soon develop the power of motility, which they retain, when kept for up to 60 days—though fertility is lost 3 or 4 weeks earlier. Between the tubules of the testis there are interstitial cells which produce male hormones. These hormones support the function of the tubules, and also pass into the blood and cause development of male sexual characters, accessory sex glands, and desire to mate. The level of interstitial cell activity required for sperm formation is less than that needed for full sexual development, and tubule damage that prevents sperm formation can occur without stopping hormone secretion. Thus there can be loss of fertility without loss of ability to mate, and vice versa. In most mammals the testes descend into the scrotum, and a cryptorchid testis, retained in the abdomen at deep body temperature, fails to produce sperm. There is a heat-exchange device in the spermatic cord; the arteries supplying the testis are in close contact with a plexus of veins in which the blood is cooler because of heat dissipation by the scrotum (Fig. 1.1).

At mating, muscular contractions carry sperm up the vas deferens (Fig. 1.1) from the epididymis, and muscle in the accessory glands expresses their secretions, into the muscular part of the urethra. Thence the semen is projected through the penis and ejaculated.

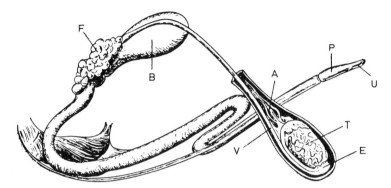

Fig. 1.1 Reproductive organs of the bull. **A,** Blood vessels in scrotal cord supplying testis. **T,** Testis; forms spermatozoa. **E,** Epididymis; stores spermatozoa. **V,** Vas deferens; conducts spermatozoa in ejaculation. **F,** Vesicula seminalis; supplies fluids diluting spermatozoa on ejaculation. **P,** penis; intromittent organ. **U,** opening of urethra; conducts semen, and also urine from the bladder. **B,** Bladder; stores urine. (Marshall, F. H. A. and Hammond, J. (1952). *Fertility and Animal Breeding*. Ministry of Agriculture Bulletin No. 39.)

Only a small fraction of the ejaculate comes from the epididymis. The accessory gland secretions add bulk, and also stimulate intense motility in the spermatozoa — though this is not an essential function because sperm taken from the epididymis will develop motility spontaneously. The amount of accessory secretions, and degree of dilution of the sperm, vary greatly in the different species of farm animals. Table 1.1, in very round figures, illustrates this; actual volume and composition of an ejaculate depend upon breed, and upon the conditions of collection. In the ruminants the semen is rather concentrated; it is sprayed over the opening of the cervix and the anterior end of the vagina (Fig. 1.2) and only a little of

Table 1.1 Size of ejaculate, and typical sperm density and numbers, of various species.

Species	Volume (ml)	Density ($\times 10^6$ ml^{-1})	Total sperm number ($\times 10^9$)
Jackass	20–80	450	20
Stallion	50–250	120	10
Boar	150–500	100	25
Bull	2–8	1000	4
Buffalo	1.5–6	1000	3
Ram	0.3–1.6	3000	3
Goat	0.3–1.6	3000	3
Cock	0.2–1.0	3500	3

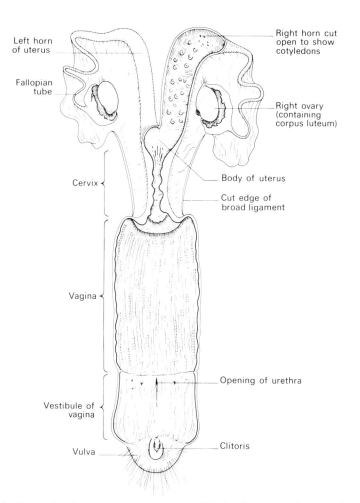

Left horn of uterus

Fallopian tube

Right horn cut open to show cotyledons

Right ovary (containing corpus luteum)

Cervix

Body of uterus

Cut edge of broad ligament

Vagina

Opening of urethra

Vestibule of vagina

Vulva

Clitoris

Fig. 1.2 Reproductive organs of the cow, slightly diagrammatic, seen from above. The ligament that supports the uterus in the body cavity has been cut away at each side and the vagina cut open in the mid-dorsal line. The cut extends through the folds of the cervix, and along the right horn of the uterus, to expose the cotyledons in its lining (the foetal membranes fuse with the cotyledons to form the placenta).

it normally reaches the uterus. The boar and the stallion ejaculate a large volume directly through the cervix into the uterus, which rapidly concentrates the semen by absorbing most of the accessory secretions.

During oestrus or heat (the only time at which mating is allowed) the mucus in the cervix is thin and watery, and provides a medium in which sperm can swim freely. The uterine muscle at oestrus shows strong rhythmic activity and this activity is reinforced when the vagina is distended at mating or insemination: there is reflex discharge of a hormone (oxytocin) from the posterior pituitary gland and this hormone is carried thence in the bloodstream, and so reaches the uterus and augments its activity.

Some sperm are carried to the site of fertilization, the top of the Fallopian tube (Fig. 1.2), within minutes of insemination—largely by the activity of the uterus and tubes. Sperm are not, however, at once capable of fertilizing an egg, but have first to undergo a ripening process (capacitation) which requires a period of (in the rabbit) some hours in the uterus or oviduct.

Ovulation, the release of the egg or eggs from the ovary, occurs in most species near the end of the period of oestrus. The egg cells develop in the ovary within structures known as follicles. The egg is immediately surrounded by a membrane, the zona pellucida, which corresponds to the membrane surrounding the yolk of a bird's egg. The egg is surrounded by cells, the granulosa layer of the follicle, and outside this again are blood vessels and cells which secrete female hormone (Fig. 1.3).

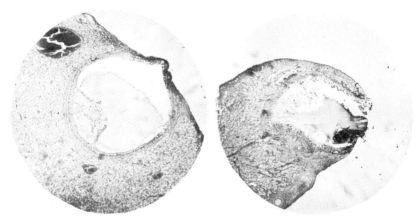

Fig. 1.3 Section through a ripe follicle containing the egg and liquor in which it floats (*left*), and one which has just ruptured (*right*) and discharged the egg; round its walls are the granulosa cells which will later form the corpus luteum. (Hammond, J. and Marshall, F. H. A. (1925). *Reproduction in the Rabbit.* Edinburgh.)

As the follicles grow, rather viscous fluid (the liquor folliculi) accumulates among the granulosa cells. At ovulation this fluid washes the egg, surrounded by zona and granulosa cells, into the top of the Fallopian tube. From the emptied follicle there then develops a solid structure, the corpus luteum, mainly by great enlargement of the granulosa cells which lined the cavity of the follicle, and by the ingrowth of blood vessels from the follicle wall.

In most species the egg is shed in an unstable condition of arrested cell division, which is resumed when the sperm head enters the egg; if not soon fertilized the ovum becomes incapable of normal development. The sperm head has, in front of the nucleus, a compartment (the acrosome) containing enzymes. By the release of these enzymes the sperm penetrates between the cells surrounding the egg, and through the zona. Having shed the outer acrosome membrane the cytoplasmic membrane surrounding the nucleus then fuses with that of the egg cell. After fertilization, penetration of further sperm is resisted. The process of capacitation seems to be a partial destabilization of the acrosome membrane. Whether fertilized or not, the egg is transported down the oviduct and is normally transferred to the uterus about three days after ovulation, at which time the uterine secretions can supply its needs.

Fertility

Apart from identical twinning, in which a single egg gives rise to more than one young, the number of young born can be no more than the number of eggs shed at the heat period preceding conception. The number of eggs shed is determined by the activity of the animal's pituitary (discussed on pages 15–17). However, there can be failure of eggs to be fertilized, and failure of fertilized eggs to complete development (Fig. 1.4).

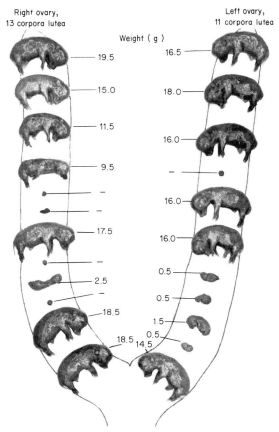

Fig. 1.4   The foetuses from a sow showing their relative position in the uterus and weight of each foetus. Although pregnancy is not far advanced, many of the foetuses have degenerated and others are about to die. (Hammond, J. (1914). *Journal of Agricultural Science*, **6**, 263.)

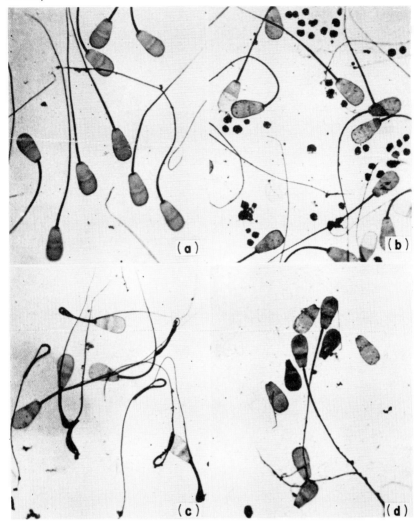

Fig. 1.5 Sperm taken from a bull at successive intervals from a period of full fertility to one of sterility. **(a)** Normal sperm; **(b)** with many free protoplasmic droplets; **(c)** with bent tails; **(d)** with heads and tails separated. (Lagerlof, N. (1934). *Acta Pathologica et Microbiologica Scandinavica*, Supplement 19.)

The semen inseminated must contain numerous sound sperm. Although only one sperm is required to fertilize an egg, the chances of any one sperm reaching an egg are very small. Production by the male of few, or defective, sperm is not uncommon (Fig. 1.5). The time at which mating is made in relation to the time of ovulation is also important. The sperm cannot live very long in the female tract (the period varies considerably between species) nor does the egg long remain capable of fertilization.

Survival of fertilized eggs depends on their adequate nutrition, and this in turn depends on the maternal hormones acting on the uterus. Defective genetic constitution of the fertilized egg may lead to its death in the course of development. Diseases, such as contagious abortion, which affect the uterus, also cause loss of foetuses.

Artificial insemination and inovulation

The procedures for storage of semen of the various species have been worked out largely by trial and error, and details are given in the relevant following chapters; here only some general comment is intended.

When using fresh semen, it is normally sufficient to assess the proportion of sperm showing good motility, and to ensure that a sufficient number of these is inseminated. However, motility is a property of the sperm tail, and fertilizing capacity one of its head: highly motile stored sperm are not necessarily fertile.

Motility is assessed at body temperature, because it is depressed on cooling; cooling is therefore employed for storage, with the object of delaying consumption of the sperm's presumably limited vital reserves. In addition a diluent is added, to give a suitable concentration for insemination, to supply a source of energy (usually glucose or fructose) for the sperm, and perhaps also to neutralize toxic products of sperm metabolism; antibiotics may also be added.

In species where it has been possible to preserve sperm by deep freezing, the method employed is to replace some of the water in the sperm by an agent, usually glycerol or dimethylsulphoxide, which seems to act in much the same way as gelatin added to an ice-cream mix, limiting the size of ice crystals formed and thus also mechanical damage to the sperm. The sperm suspension supercools, and freezing is carried out rapidly to minimize the temperature rise occurring with release of the latent heat of crystallization.

It is not understood why semen of some bulls withstands freezing better than that of others, nor why excessive dilution or cooling of semen should damage sperm. Addition of milk or egg yolk to saline dilutors may provide some protection from this damage, perhaps by an action on the sperm surface of some organic constituent. It is likely also that cooling affects the normal concentration differences of various ions across the cell membranes.

The pioneering work on the procedures for handling, storing and transferring developing embryos was done by Pincus and Chang with laboratory animals, and on freezing fertilized eggs by Whittingham. Eggs will continue to develop if incubated at body temperature in blood serum (which may need to be heat-treated). They can also be maintained in balanced salt solutions if simple nutrients and some serum albumen is added. Eggs of other species are adequately warmed and nourished if transferred to the rabbit Fallopian tube (the tube must be tied off to prevent the eggs being passed into the uterus); so the rabbit can be used as a convenient portable incubator.

Successful development of a transferred egg requires that the post-ovulation maturation of the recipient female tract be closely matched to that of the egg, which must interact with it to form a placenta (see below). Thus, either the time of ovulation of donor and recipient animal must be synchronized, or the development of the egg must be suspended until a recipient is available at the appropriate stage.

The uterine secretions that will nourish the egg will also support the growth of a variety of bacteria. Transfer therefore requires precautions against infection; also the volume of fluid introduced with the egg must be minimal, for the uterus will tend to expel it—and with it, perhaps, the egg.

Pregnancy

There are three successive phases of nourishment of the developing young during pregnancy; their relative durations, and the extent to which they overlap, vary with the different species. In the first phase the developing egg depends on the fluid surrounding it—the secretions of the oviduct and of the glands of the uterus. The nutritional needs of the egg change as it develops and the earlier tubal stages of development are not supported by the uterine secretions.

Within the cavity of the uterus the dividing cells of the egg, at first still within the zona, arrange themselves in a fluid-filled sphere, the blastocyst. By pumping fluid into itself, the blastocyst expands and so makes firm contact with the lining of the uterus.

In the second phase the placenta is formed by interaction between maternal and embryonic tissues. Cells of the embryonic membranes attack and digest the uterine lining, and uterine cells proliferate and so provide further food for the embryo.

In the third phase, transfer of material from mother to foetus is passive rather than active. The rate of maternal blood flow through the uterine lining, and its vascularity, are much enhanced in the areas of uterus which have responded to embryonic attack. Foetal blood vessels in the placenta, which have developed to carry the products of uterine digestion from the placenta to the body of the growing foetus, come into closer contact with the maternal vessels as intervening tissues are removed. Thus the foetus can obtain nourishment by direct diffusion of material between the two bloodstreams. The nature of the attachment formed varies between species. In some, relatively large molecules such as antibodies, formed by the mother in response to infections, can pass to the foetus *in utero* and so provide some passive immunity to the young. In farm animals such immunity has to be acquired from the colostrum in the first day or two of postnatal life.

During pregnancy the uterus grows, and its muscle fibres enlarge, under the combined influences of distention by the foetal fluids and of changes in the levels of circulating hormones. In the glands of the cervix, viscous mucus accumulates and forms a barrier between uterine and vaginal contents (Fig. 4.20). Growth occurs not only in the uterus but also in the

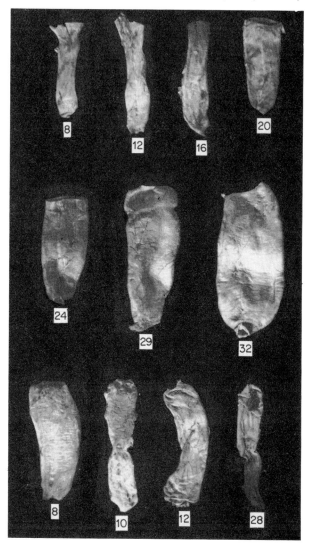

Fig. 1.6 Changes which take place in the development of the vagina of the rabbit from the 8th to the 32nd day of pregnancy, and (*bottom row*) its involution from the 8th to 28th day after parturition. The numbers show the days pregnant or days post-partum.

vagina (Fig. 1.6) and in the mammary glands. These changes are brought about under the influence of the hormones of pregnancy. The growth and dilation of the vagina are necessary for the passage of the young at birth. This vaginal development mainly takes place rather late in pregnancy, and if it is inadequate leads to difficulty at parturition—and similarly with the relaxation of the connective tissue of the cervix.

In birds, no corpus luteum develops in the ovary when the yolk is shed, and formation of the rest of the egg is completed as it passes through the oviduct, to be laid about 24 hours later. The bird can produce an egg every day. In higher mammals a corpus luteum is formed and while it is active no other egg matures, and the animal does not come on heat again. The corpus luteum secretes a hormone (progesterone) which—even if the egg is not fertilized—brings about (though often on a small scale) many of the changes of pregnancy. Rhythmic activity of the uterus is suppressed, the uterine glands are modified so that the egg, if fertilized, can live in the fluid they form, and the uterus is enabled to form a placenta if stimulated by the foetal membranes; also the mammary glands are caused to develop.

The probable course of evolution of pregnancy has been first the appearance of the corpus luteum, with retention of the egg for incubation within the oviduct while the corpus persisted in the ovary; then the reduction of egg size and development of placentation; and then the prolongation of pregnancy so that the young shall be more mature at birth. This prolongation is achieved differently in different species.

The hormones of the placenta may, to a variable extent, supplement or replace those of the corpus luteum and other ovarian structures; in some species the placenta produces hormones specifically affecting the maternal ovary.

In many species there is an arrangement whereby the uterus, if it does not contain developing eggs, cuts short the life of the corpora lutea, so that the animal can come on heat and ovulate again. Through this mechanism the foetus may, indirectly, bring its own gestation to an end (see p. 18).

The reproductive hormones

Progesterone, the female hormones (oestrogens), the male hormones (androgens)—and the hormones of the adrenal cortex also—are all steroids; that is to say, they are all chemically similar, and can be derived by modification of cholesterol. Progesterone, indeed, may be an intermediate compound in the formation of the sex hormones, and androgens precursors of oestrogens. Testosterone is generally the principal, and most potent, androgen secreted by the testis; but many of its effects are produced through its subsequent conversion to dihydrotestosterone. An effect of androgens is the promotion of muscle growth; but this is an effect produced by some substances that in other respects are rather weak androgens. Such natural steroids, and synthetic ones with the same effect, are commonly called 'anabolic steroids'. Anabolism, strictly speaking, refers only to increase of body mass; progesterone (see Fig. 2.3) can also cause weight increase, but is not classed as an anabolic steroid.

Male hormones are not exclusively male, nor oestrogens female; for example, the reddening of the comb of the pullet, as it comes into lay, is due to androgen from the ovary, and there is androgen in female human urine (mostly derived from modification of adrenal hormones).

Many sexual functions are affected by more than one type of steroid hormone; in some, a particular effect is obtainable only by a combination of hormones, and the effect produced may depend upon the proportion in which the hormones are present. The development of uterine glands by progesterone is enhanced by a small amount of oestrogen, but is antagonized by oestrogen in excess.

The steroid hormones are passed into the blood as fast as they are formed, but are quickly inactivated; about half of what is released at any moment will have been destroyed half an hour later. Some of the steroid in the blood is in simple solution, most of the remainder is loosely bonded to blood proteins; the extent of bonding can affect the time the hormone remains in circulation, and perhaps its relative availability to different organs. There are also plasma proteins with high affinity for particular steroids. (Such proteins can be used to measure minute quantities of that steroid extracted from body fluids; if steroid chemically labelled with a radioactive isotope is added, and then the protein, the radioactivity binding with the protein will depend upon the proportions of labelled and unlabelled steroid present.)

The natural steroid hormones are nearly inactive when taken by mouth; partly they are destroyed in the gut before absorption, partly inactivated afterwards, but before they reach the general circulation. They can be administered by injection in oil (they are not very soluble in water), when they will pass gradually—but in decreasing amount—into solution in the body fluids, and so remain in circulation much longer than if added directly to the blood. A convenient method of obtaining steady circulating levels for long periods is to implant pellets of the substance under the skin. The rate at which they dissolve then depends on the surface area of the tablets and the solubility of the particular substance. Instead of a pellet, an implant can be made of crystals contained within a capsule of a plastic in which the substance is soluble. Absorption in aqueous solution through the vaginal wall is obtainable at an adequate rate by increasing the surface area for solution (see Fig. 5.2).

Substances with oestrogenic activity are found in some plants, including clovers. They may be present in sufficient amount to cause mammary development in wethers and sterility in ewes. There are many substances— of which diethylstilboestrol is perhaps the best known—which have oestrogenic activity but are of simpler chemical structure than the steroids.

Stilboestrol is of importance because it is cheaper to make than the natural hormones; in some ways it is more potent than any natural oestrogen. Potency is perhaps partly due to the firmness with which a substance binds to its site of action: potency is also enhanced if the compound is less easily inactivated or excreted by the animal—it will remain longer in the circulation. There are a variety of synthetic steroids which have hormone activity and which can be given by mouth. Their chemical structures are based on those of the hormones, altered in a way that interferes with the natural methods of inactivation; consequently they also tend to be persistent in action. There are also substances (such as the anti-androgen cyproterone) which block hormone action, though not themselves possessing activity—or only very weakly.

From the essential polyunsaturated fatty acid, arachidonic acid, a group of transmitter substances collectively known as prostaglandins (PGs for short) are derived. As with the steroid hormones, differences of structure are associated with different specificities of action. It seems that the drug aspirin acts by interfering with their formation. They are widely distributed in the body, but in general their effects seem to be purely local, as they are rather unstable and rapidly inactivated, and in ordinary concentrations do not survive passage through the lungs.

One of the prostaglandins (PG $F_{2\alpha}$)provides the means by which the non-pregnant uterus cuts short the function of the corpus luteum. It is formed by the lining of the uterus and passes in the uterine vein to the ovarian artery by diffusion through the blood vessel walls, in a plexus of vessels analogous to that which cools the blood supply to the testis (p. 4). Several stable analogues of this protaglandin have been synthesized.

The activities of the gonads (ovaries and testes)—both in producing hormones and in maturing germ cells—are regulated by hormones (gonadotrophins) of the anterior pituitary gland. The pituitary lies at the base of the brain, a little behind the point at which the nerves from the eyes enter. The posterior pituitary is in fact part of the brain, and the hormones of the posterior pituitary are formed by nerve cells whose bodies lie in that part of the brain (the hypothalamus) immediately above the pituitary.

The anterior pituitary itself is largely regulated by centres (groups of nerve cells) in the hypothalamus. These centres integrate the various influences—age, nutrition, photoperiod, stress, steroid hormones and emotion—which affect gonad function. The regulation is by hormones (analogous to those of the posterior pituitary) carried in blood vessels running directly from the hypothalamus to the anterior pituitary. Also in the hypothalamus are centres concerned with appetite, water balance, body-temperature regulation, and emotional adjustments of a variety of functions, including sexual behaviour.

The anterior pituitary produces two gonadotrophins (FSH and LH) and a third hormone, prolactin, which is also particularly concerned with reproductive functions. Prolactin is also known as the lactogenic hormone. The earlier stages of mammary development are regulated by the ovarian steroids, but prolactin is necessary for full mammary growth and for secretion of milk. In some species it has a role in regulating the function of the corpus luteum. That it has other functions also is indicated by its presence in the pituitary of males, and of sub-mammalian animals.

FSH and LH signify respectively follicle-stimulating and luteinizing hormone. FSH is especially concerned with germ cells (i.e. with follicles and the testis tubules), and LH with steroid hormone secretion. But the full development of follicles requires steroid hormones, and follicle growth requires both FSH and LH. Release of the egg and formation of the corpus luteum particularly require LH.

The pituitary is regulated by the brain. Nervous transmitter substances in general are modified amino acids or peptides, and those (which have so far been identified) controlling the anterior pituitary are composed of relatively few amino acids. Some of these hypothalamic transmitters stimulate, and some inhibit, hormone secretion. It appears that a single transmitter substance, known as LH-RF (RF signifying releasing factor), controls the secretion of both FSH and LH. It has been identified, and synthesized, and active analogues have also been prepared. The extent to which FSH or LH secretion is stimulated depends partly on the steroid hormones simultaneously acting on the pituitary cells. This (and other releasing and inhibitory factors) seem normally to be liberated in a pulsatile fashion rather than in a steady stream, and this may well influence the response to them of the pituitary cells. It may prove difficult to apply these substances usefully to manipulate pituitary functions.

Hormone production by the placenta is of steroids and of anterior pituitary-like substances—but this is not the same in all species. In all cases their origin seems to be in the foetal membranes in contact with the uterus. Progesterone and oestrogen are formed there from steroid precursors supplied by the foetus or dam. A placental 'lactogenic' hormone is formed in ruminants—probably by foetal cells that migrate into the uterus—and gonadotrophins are secreted in the mare and the human.

In artificially stimulating follicle growth and ovulation, it is not necessary to have pure FSH and LH. The proportions of these hormones are different in the pituitaries of different species—that of the mare, for example, being relatively rich in FSH, and that of cattle in LH. However, pituitary glands are small, and not easily accessible: pituitary hormones are therefore scarce, and expensive. Much more plentiful, and more easily obtained, are two placental gonadotrophins, MSG and HCG, which can be used as substitutes for the pituitary hormones.

MSG, the mare serum gonadotrophin, is present in the blood in very large amounts during early pregnancy in the mare (see Fig. 3.4). It is a hormone that combines the activity of both FSH and LH, and is predominantly follicle stimulating in its effect. It does not pass into the urine, and so is more persistent in action than pituitary gonadotrophins. HCG (human chorionic gonadotrophin) appears in the urine quite soon after conception and is later present in large amounts. Its capacity to cause ovulation in the rabbit is the basis of the Friedman pregnancy test, and HCG can be used as a substitute for LH.

The gonadotrophins are glycoproteins, large molecules, and with species differences of composition. Injected into animals of a different species, they are capable of inducing formation of antibodies that will neutralize them if treatment is repeated. Such antihormones are of some experimental use; at least when MSG is given to sheep and cattle they seem to be of little practical importance.

Control of ovulation

Rabbits (and various other mammals) normally ovulate only after sexual excitement. The number of ripe follicles that develop in the ovaries before mating—and hence the number of eggs shed when the rabbit mates—is regulated by the level of gonadotrophins secreted by the pituitary before mating. This level is determined by a negative feedback mechanism: oestrogen produced by the follicle depresses release from the pituitary of the gonadotrophins causing follicle development.

The active corpus luteum through its progesterone secretion, also has a negative feedback effect on follicle growth. When the corpus luteum of the cow is removed (Fig. 4.3) or is caused to regress by treatment with prostaglandin the cow comes on heat, and ovulates again, within a few days. Conversely, if the cow is treated continuously with progesterone, or the corpus luteum persists—as in pregnancy—follicles do not ripen and ovulate.

The cow, and most mammals, the chicken, and a few other birds, ovulate spontaneously. In them the primitive device of the ovulating discharge of gonadotrophin being released by sexual excitement has been short-circuited—in different ways in different species. In the rat, the inbuilt biological clock which regulates the animal's diurnal pattern of activity and rest causes ovulation always to be at the same time of day. In the fowl, time of ovulation is such that eggs are laid always in daylight hours; the mechanism seems to be an interaction between a clock and a steroid feedback from the ovary.

In the sheep, cow, and others, it would appear that oestrogen feedback from the ripe follicle triggers the ovulating discharge from the pituitary: injected oestrogen can lead to ovulation, but progesterone antagonizes this oestrogen effect. The number of follicles normally developing at oestrus in such animals would then seem to be determined by combined oestrogen and progesterone feedback as the corpus luteum regresses. Breed differences in fertility (see Table 5.1), so far as they reflect ovulation rate, would then be due to genetic differences in sensitivity at some point of the feedback mechanism.

In experiments with laboratory animals it has been shown that the adult male hypothalamus lacks this trigger mechanism, but retains it if the testes are removed at birth. Conversely, the ovulating centre is destroyed in the female if it is treated with androgen (or a large dose of oestrogen) but only if it is given during a critical period soon after birth.

The sensitivity of the gonadal feedback is not constant, but changes with ageing—decreasing at puberty to permit sperm formation and ovulation. In animals with a restricted breeding season there is seasonal change of sensitivity. Breeding seasons may be determined by a variety of factors (Marshall, 1936); sometimes food supply is the major factor, often photoperiod (length of day) is responsible. Sheep and deer, which are autumn or winter breeding, are said to be 'short-day' animals, while ferrets, poultry and horses are stimulated to breed by long photoperiod.

However, response to daylength is not so simple as these statements might seem to imply. In the 'long-day' ferret, the onset of the breeding season in the spring is as much a delayed response to short day as an immediate one to lengthening daylight hours; very long photoperiods are less stimulating than shorter ones; and females kept for a long period on 14-hour photoperiods may fail to breed, though they will do so when kept on short days.

Control of parturition

During pregnancy many functions develop, all to reach maturity at parturition. The mammary gland, cervix and vagina grow; the placenta ages so that it will separate from the uterus; in the foetus the gut, the lungs and the kidneys, and skin insulation and temperature regulating systems all mature, so as to function at birth.

The events of the ovarian cycle, and subsequent heat, in the non-pregnant mammal provide a miniature of pregnancy. In the rabbit and ferret, which ovulate only after sexual excitement, mammary development is such that the resemblance is obvious and non-pregnant mated females are said to be pseudopregnant. As in other species, with the decline of the corpus luteum and the development of large follicles, the cervix re-opens, strong rhythmic contractions return to the uterine muscle and the cells lining the uterus are replaced; in the human, the shedding of the uterine lining is akin to the separation of the placenta.

Pregnancy has evolved differently in different groups. In the simplest case, where the life of the corpus luteum is unaffected by conception, this may well determine the duration of pregnancy. In the mare (see Fig. 3.5) where the corpus luteum does not persist, or in the cow, in which it can be removed in later pregnancy, clearly this is not the case.

The observation that, following surgical removal of the monkey foetus, the placenta was delivered at about the normal time for birth, indicates that mother or placenta determine that time. Decreased sensitivity of uterine muscle, after ovulation, to the posterior pituitary hormone oxytocin, and increasing sensitivity towards the end of pregnancy, suggest either an oestrogen–progesterone balance mechanism in regulating uterine excitability, or else a role for the posterior pituitary in initiating parturition. The use of oxytocin as a drug to induce labour in women supported that idea. However the role of oxytocin is now seen rather as one of reinforcing uterine activity following dilation of the cervix.

In the mare, where the sire of the foal clearly affects the length of gestation (p. 51), obviously foetal tissue decides the time of birth. The clue to the mechanism came when it was observed that cows carrying calves with a recessive genetic defect had very prolonged pregnancies. The defect was due to lack of anterior pituitary development in the calf; and experiments with sheep showed that parturition is initiated by the foetal pituitary. Secretion of the pituitary hormone ACTH (adrenocorticotrophic hormone) stimulates the foetal adrenal cortex; the resulting adrenal (glucocorticoid) hormone then acts on the placenta to change the pattern of its hormone production. Oestrogen is formed from progesterone, and this in turn seems to cause the uterus to produce prostaglandin, and thus to terminate function of the corpus luteum.

Possible future developments

For large scale application of ovum transfer, the method must be simple, non-surgical, and fertilized eggs be plentifully available. The technique of ovum transfer has been used to show that eggs shed after gonadotrophin treatment—in much larger numbers than is normal for the species—are normal, and capable of development if transferred to several other females (Fig. 1.7). However, they would not all develop in the mother (see Fig. 5.4), and loss seems to start in the Fallopian tube, quite soon after ovulation.

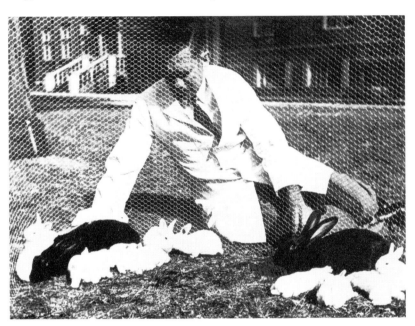

Fig. 1.7 Black host does and their white young from the transplanted ova from one superovulated white doe. In this experiment 27 ova at the 32 to 64 cell stage (63 hour) were transplanted into four black does, and 22 young were born. (Dowling, D. F. (1949). *Journal of Agricultural Science,* **39,** 374.)

There seems then to be a choice between repeated recoveries by flushing from the uterus, for transfer to the uterus of the recipient (immediately or after storage), but with considerable losses; or of recovery at slaughter, soon after fertilization, and subsequent culture until the eggs are mature enough for transfer to the uterus.

With laboratory animals it has proved possible to capacitate sperm, and to fertilize eggs, outside the body. This may be a useful way of obtaining eggs for transfer. However it will probably be a long time before the immature egg cells in the ovary can be grown to full size in culture; and it is not yet possible to bring full sized eggs to readiness for fertilization, except by treatment with gonadotrophin while still within the follicle. It will be a long time before a cow can have anything like as many offspring as can a bull.

The main purpose of using ovum transfer will be to obtain better animals than could be obtained simply by insemination of the female—better dairy, or better beef, than can be obtained with a good bull on a female indifferent for either purpose. But it should not be forgotten that the host female will supply the environment in which the genetically superior calf develops—and the early effects of environment may well be the most important.

There have been several claims of methods for separating the sperm that will give male and female offspring; so far, all seem to have rested upon results that could not be repeated, but eventually it may be possible to ensure that only heifer calves are conceived. A much more likely way of ensuring that only male, or female, young are born is by transfer of fertilized eggs whose sex has been determined. This may be done by examining either the chromosomes (p. 198) in dividing cells (taken from foetal membrane) or (sometimes) less laboriously by looking at resting nuclei; in these a special body, thought to represent the inactivated material of the second X chromosome, may be distinguishable in females.

There has been little effort to develop artificial lactation (p. 83); the value of the calf as a by-product would be lacking. Perhaps it could be useful if applied to males. In fact, when tried on a freemartin heifer the milk yield was very small. The reason was probably that male hormone acting on the mammary gland at a critical stage during foetal life restricts its capacity for future development. Experimentally it has been shown, in the rat, that after treatment of the foetus with the anti-androgen cyproterone (and castration when adult) steroid treatment produced as great mammary development in the male as in the female.

Further reading

AUSTIN, C. R. and SHORT, R. V. (Eds) (1972, vols 1–5; 1979, vol. 7). *Reproduction in Mammals.* Cambridge University Press.

COLE, H. H. and CUPPS, P. T. (Eds) (1977). *Reproduction in Domestic Animals*, 3rd edition. Academic Press, New York and London.

MARSHALL, F. H. A. (1936). Sexual periodicity and the causes that determine it. *Philosophical Transactions of the Royal Society*, B, **226**, 423.

MAULE, J. P. (Ed.) (1962). *Semen of Animals and Artificial Insemination.* Commonwealth Agricultural Bureaux, Farnham Royal (Technical Publication No. 15).

2 General principles— metabolism and growth

Growth, at least to the stage of sexual maturity, constitutes part of the process of reproduction; it involves differentiation of organs, changes of size and of body proportions, and ageing changes in function.

Metabolic precursors

Apart from minerals and trace nutrients, an animal requires protein and energy from its diet. Protein is required for maintenance, making good the breakdown of tissue associated with normal functioning, for growth, and for production of wool, milk and eggs. Energy is needed for the functioning of all cells, for the work of muscular contraction, and to generate the heat which maintains the body temperature. Energy is also needed for the process of secretion, and for the formation of new tissue.

The major constituents of the diet are protein, carbohydrate and fat. Proteins are composed of, and digested to, a variety of amino acids. Some of these are interchangeable in the animal body, others are 'essential'— that is, they have to be supplied in the feed (and are on average relatively more abundant in animal than in plant proteins). Absorbed amino acids, if not incorporated into body tissue, are quite rapidly deaminated, the ammonia being converted into urea and excreted, and the residues, according to their nature, treated by the body either as fat or carbohydrate, and stored as such, or oxidized to provide carbon dioxide, water and energy.

Dietary and body fat consists largely of a mixture of triglycerides, compounds of glycerol with long, straight-chain, fatty acids with an even number of carbon atoms. In rapid fattening the composition of body fat may be affected by the fatty acid composition of the diet, but generally the fat laid down is modified according to the species of animal and the position in which it is deposited.

The main carbohydrates of the diet are starch and cellulose; but cellulose is not directly digestible by animals. Starch is a polymer of the six-carbon sugar glucose (to which it is hydrolysed in the intestine) and glucose is the form in which carbohydrate is generally available to the tissues; but it is stored in the body as another polymer, glycogen, which is formed in the liver and in muscle. However, the amount of energy which can be stored as carbohydrate is very limited—roughly one day's supply—and storage of greater amounts is effected by the conversion of carbohydrate to fat.

As well as by oxidation, muscle can derive energy anaerobically from glycogen by splitting it to the three-carbon compound, lactic acid. The lactic acid is then either oxidized elsewhere, or is reconverted to glucose in the liver. Post-mortem, also, muscle glycogen is converted to lactic acid, and the resultant acidity is important for the development of tenderness in the muscle during hanging and cooking. Genetic selection of pigs for low body-fat content has led to excessive glycogen storage in muscle. Muscle protein normally holds water bound to it, but carcases of the affected pigs have 'watery pork', or so-called muscular degeneration (p. 263) in which muscle protein structure is altered by the high level of acidity which develops post-mortem.

Though the body has reserve energy stores of both glycogen and triglycerides, glucose and fatty acids are not completely alternative sources of energy. Some tissues, notably the brain, require a supply of glucose. Furthermore, the breakdown of fatty acids requires a supply of carbohydrate; if this is not available the two-carbon residues to which the fatty acids are reduced combine to form toxic quantities of 'ketone bodies' (acetoacetic acid and β-hydroxybutyric acid) which impair nervous function—as in human diabetes, ketosis of dairy cows, and twin-lamb disease (p. 124). For the secretion of milk the mammary gland requires fatty acids as precursors of milk fat, and glucose, both as a precursor of milk sugar and as an energy source for secretion.

Because of the limited body stores of carbohydrate, and the need for some carbohydrate to support the oxidation of fat and the functioning of tissues such as the brain, it follows that if food is withheld body protein must be broken down in quantities greater than are used in normal body functioning. This is necessary to provide carbohydrate from some of the constituent amino acids of the proteins (a process known as gluconeogenesis).

Cellulose, the other main feed carbohydrate, is only indirectly digestible—by the plant enzymes of the bacteria in the gut. This is a relatively slow and relatively inefficient process. In the chicken, where the gut capacity is relatively small, or in the young pig where appetite is high, average gut transit time of digesta is necessarily short, and energy obtained from cellulose digestion is negligible. Time, and suitable conditions, for bacterial fermentation of cellulose are provided by the large-capacity hind-gut (particularly, for example, of the horse) and the fore-stomachs of the ruminant.

Digestion and absorption of the other energetic constituents of the diet are largely completed in the fore-gut; in the rumen these other elements of the diet, as well as cellulose, are exposed to bacterial action.

Rumen bacteria derive energy from cellulose by breaking it down anaerobically, the main products being acetic, propionic and butyric acids. The three-carbon propionic acid is treated by the body much as carbohydrate, while the two- and four-carbon fatty acids are fat precursors, for example of milk fat. The accumulation of hydrogen ions in the rumen, which would otherwise inhibit further fermentation, is prevented by a constant flow of alkaline saliva—a flow augmented when the animal eats or ruminates.

Other rumen bacteria are capable of splitting starch to lactic acid, as muscle can glycogen. The resulting acidity may suppress the lysis of cellulose, and consequently reduce the availability to the animal of the two- and four-carbon compounds.

Feed protein entering the rumen is largely deaminated, and the resulting ammonia used by the multiplying bacteria to synthesize bacterial protein; urea secreted in the saliva can be converted to bacterial protein if energy is also available.

Experimentally, the provision of high quality protein which has by-passed the rumen can result in increased yields of milk or of wool, So it would seem that either the availability of one or more essential amino acids limits the rate of production, or else the level of such amino acids in the blood affects the secretion of hormones involved in regulating these processes.

Metabolic rate

First claim upon available food supplies goes to maintenance of the animal's body temperature. Some heat is inevitably lost by evaporation of water through the skin and lungs: unless the surroundings are very warm there is also conductive loss from the body surface. Rate of air movement greatly affects rate of loss, which (except by radiation) will be proportional to the temperature difference between the body surface and the air.

The vital processes, including movement and digestion of food, entail heat production: rate of resting heat production can be varied, in the long term, through the pituitary and thyroid glands. Within a range of temperatures—the lower of which is usually known as the critical temperature—heat production and loss are balanced by altering skin temperature (mostly at the extremities) by altering the blood flow.

In ruminants, heat production is much affected by level of gut bacterial activity. Figure 2.1 shows how plane of nutrition affects basal heat production and critical temperature: and also how heat production, and loss, increase linearly when environmental temperature falls below the critical value. Because of their larger surface area, and relative lack of insulation, critical temperature is higher in younger animals. Below the critical temperature, food otherwise available for growth has to be burnt to maintain life (see Fig. 10.14).

When sufficient heat cannot be lost from the skin surface (including loss by sweating in those species able to sweat effectively) evaporation is increased by rapid shallow breathing (see Fig. 10.9)—though this entails extra heat production. Further adjustment involves cutting down heat production, both by reducing activity and by reduction of appetite (see Fig. 10.10), and so of the heat generated in digestion. Diurnal fluctuation of normal body temperature may play a part in adjustment of heat pro-

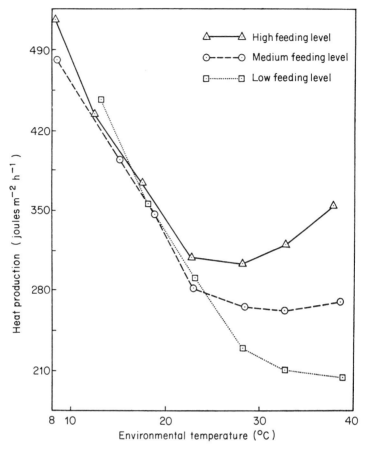

Fig. 2.1 Effect of environmental temperature on heat loss of closely shorn sheep on high, medium and low planes of nutrition. (Armstrong, D. G., Blaxter, K. L., Graham, N. M. and Wainman, F. W. (1959). *Animal Production*, **1**, 1.)

duction and loss. Skin and hair pigmentation are important in determining heat exchange by radiation.

The regulation of heat loss (and so control of body temperature) is effected by nerve centres in the hypothalamus. These are provided with messages from temperature sensors in the skin and elsewhere—including the rumen—and are themselves sensitive to the temperature of the blood flowing through them. This central 'thermostat' does not have a constant setting, but one which fluctuates diurnally; the difference between its maximum and minimum setting over the 24 hours allows for a considerable temporary imbalance between heat production and heat loss, because of the high thermal capacity of water—which constitutes 70 or 80 per cent of the body weight.

Air breathed in through the nose passes over the moist nasal mucosa; thus it is saturated with water vapour at body temperature before reaching the lungs, and the nasal mucosa is cooled by the evaporation of that water. When the moist air is expired through the nose some moisture is recovered by condensation, and the membrane correspondingly rewarmed. In many species there is a heat exchange mechanism—such as that of the testicular cord (p. 4)—and arterial blood passing to the brain is cooled by venous blood from the nose. By this device the most vital organ, the brain, is kept near its proper temperature—but at the expense of the rest of the body.

Long-term adjustment to winter temperatures is brought about in some animals by increased deposition of subcutaneous fat and growth of a denser coat of hair, so that they are better insulated from heat loss. These changes are photoperiodically regulated. In tropical cattle the coat is always short, in British breeds the winter coat is longer (see Fig. 10.13). The shedding of a hair from its follicle, and replacement by a fresh hair, is primarily regulated by ageing changes in the follicle; but these are hormonally modified through the adrenal cortex. In the ferret, adrenal cortex hormone delays shedding and prevents replacement of hairs, and oestrogen potentiates the action of the adrenal steroid. This interaction probably explains the occurrence of moulting in chickens at the end of a period of lay. The retention of winter coat in an undernourished, sick, or heat-stressed animal probably reflects increased adrenal secretion in adjustment of the pattern of metabolism—which change is also shown by body conformation effects (see Fig. 10.12).

The metabolic hormones

The pattern of metabolism and growth can be considered as the outcome of a competition between the different tissues for the nourishment provided for them in the blood; the amount and nature of that nourishment being dependant upon appetite, or upon limitations of the available diet.

Factors influencing the competition will be the relative size and inherent growth potentials of the tissues; but there are also adjustments brought about by hormones limiting the supply of nutrients to the various tissues, or limiting their capacity to assimilate them. For the survival of the individual, priority is given to the needs of the nervous system and maintenance of body temperature, ahead of requirements for reproduction and for growth.

The regulation of metabolism is not fully understood—and it may be added that the hormones involved were named when even less was known about their action. The brief account of metabolic hormones which follows is incomplete, and, in the sense that little of the knowledge available can yet be applied, perhaps unnecessary in a book such as this. Without such knowledge, the geneticist can change the growth pattern of animals, and the statistician produce formulae which describe body composition changes, or milk production rates, with fair accuracy; nevertheless, to understand growth one must understand the mechanisms involved.

Because the hypothalamus is the part of the brain concerned with the regulation of body temperature, of appetite, and of pituitary hormone secretion, it is reasonable to regard it as coordinating metabolism through its control of anterior pituitary hormone secretion.

But before itemizing these hormones, reference must be made to those of the pancreatic islets and of the adrenal gland.

Insulin, secreted by the pancreas, facilitates tissue uptake of glucose from the blood. In non-ruminants the level of blood glucose is held rather constant, and high; the pancreatic islets respond principally to an increase in the level of glucose (as after absorption of carbohydrate from the gut), increasing insulin production and thus removal of glucose from the blood. If the blood sugar falls—as when food is withheld—other mechanisms restore the glucose level. However, in ruminants absorption is a more continuous process, very little carbohydrate is usually absorbed as such, and blood glucose levels are lower and more variable—and probably less important in regulating insulin secretion.

The adrenal consists of two glands, not quite independent of each other. The medulla of the adrenal is nervous tissue and produces adrenalin and noradrenalin, which are nervous transmitter substances—both in parts of the brain and in the body (for instance in regulation of heart beat and of blood vessels). When released into the blood (mainly in emergency situations) they have the effect of mimicking, or reinforcing, the activity of the corresponding nerves. Acting on the liver, adrenalin, for instance, causes a discharge of glucose (derived from liver glycogen) into the blood; it also causes a short term increase of heat production.

As noted earlier (p. 13) the adrenal cortex produces steroid hormones: some of these, the mineralocorticoids, need not be further considered. The others, the glucocorticoids, are secreted in response to the pituitary hormone ACTH. Associated with the release of these specifically corticoid hormones there is also a release of such steroids as progesterone, oestrogens, and weak androgens. Glucocorticoids have several roles in the response of the body to infection; but they are so called from their action in promoting gluconeogenesis—the breakdown of protein to form glucose.

The anterior pituitary hormone TSH (thyroid stimulating hormone) through its control of the rate of thyroid hormone secretion, regulates the animal's resting rate of heat production, and metabolic activities of tissues generally (including milk secretion, see p. 84), but not selectively.

Removal of the pituitary early in life leads to dwarfing. Just as gonadotrophin treatment can push the ovulation rate beyond the normal genetic potential, so treatment with pituitary 'growth' hormone can push growth beyond its normal limits (see Fig. 2.2). In fact, the increased length of bone which results from treatment is not a direct effect of the growth hormone. Length growth of limb bones is by multiplication of cartilage cells between the shaft and the ends (epiphyses) of the bone; when these fuse the bone can grow only in thickness. Growth hormone affects this growth by an action through the liver, causing production of a principle acting on the epiphyseal cartilage. If the diet is deficient in protein much less of this liver hormone is produced. Growth hormone secretion does not cease when growth ceases; but a fall with ageing in the level of secretion is undoubtedly a factor in determining adult size.

Fig. 2.2 Changes wrought in the Dachshund by administration of the anterior pituitary growth hormone. The animals depicted are a pair of untreated (*above*) and treated (*below*) littermate brothers. (Evans, H. M., Meyer, R. K. and Simpson, M. E. (1933). *Memoirs of the University of California*, **2**, 423.)

It will be noted from Fig. 2.2 that the connective tissues, skin and bone, are particularly affected. If pairs of young animals are fed the same quantities of food, and one of the pair is treated with growth hormone, the carcase of the treated animal is found to contain more bone and muscle, and less fat, than that of its control. An action of growth hormone is to antagonize the action of insulin in making glucose more readily available to the tissues. As with the sex steroids, it may be supposed that synergism and antagonism between these, and other metabolic hormones, produce relatively different effects in different tissues.

Oestrogen and progesterone acting on the udder normally regulate the early stages of mammary growth; but under experimental conditions insulin may be made the limiting factor that determines gland development. The pituitary hormone prolactin (and placental lactogenic hormone) is required for, and possibly regulates, later development of the udder. Prolactin is closely related, structurally, to pituitary growth hormone—and its actions are not restricted to the mammary gland. Regulation of prolactin secretion in some species, as with gonadotrophin secretion, seems to be partly under photoperiodic control. While prolactin is required for development and maintenance of mammary tissue, it does not normally seem to be a regulator of rate of milk secretion, whereas growth hormone treatment (see Fig. 4.25) can increase yield (but perhaps through influencing the supply of precursors of secretion, rather than by a direct action on the gland).

The pituitary adrenocorticotrophic hormone (ACTH) is so called from its function of controlling adrenal secretion of glucocorticoid hormones; through them, it affects gluconeogenesis (and also a variety of responses to infection). However it also has a direct action on adipose tissue; triglyceride is hydrolysed and free fatty acids are liberated into the blood. Thus ACTH mobilizes energy reserves of both fat and carbohydrate (through gluconeogenesis).

Pituitary secretion of ACTH is under a dual control; there is a 'resting' level of activity controlled by a feedback from the adrenal (as there is for gonadotrophin from the testis, or TSH from the thyroid), and a 'stress' or emergency discharge which can be induced nervously or by adrenalin. At oestrus, with raised blood level of oestrogen there is also a raised blood level of glucocorticoids; so it may be that the increased excitability and decreased appetite of the animal on heat is associated with an increased level of ACTH secretion. Pituitary extracts, as well as containing ACTH, also contain a much larger molecule which incorporates ACTH, a hormone (MSH) which affects skin pigment cells, and a principle which blocks transmission in nervous pathways conveying the sensation of pain. It is not known whether release of these substances is independently regulated; the seasonal coat colour change of some species seems to indicate a photoperiodic control of MSH (and perhaps ACTH) secretion in them.

Appetite

Growth is clearly dependent upon food intake, and this is governed by appetite, which is regulated by nerve centres in the hypothalamus. Normally appetite is high in the young animal and falls as adult size is approached, but surgical or chemical (by gold thioglucose) destruction of the 'satiety' centre results in maintained appetite until the fat depots are replete, and the animal has become obese. Appetite fails in a variety of specific nutritional deficiency conditions: normally it is believed to be regulated by the level of nutrients available to the 'satiety' centre.

The high growth rate of the young mammal depends on the supply of plentiful and readily digested food—the milk (Fig. 7.12). Within limits, dilution of food with inert material soon leads to a compensatory increase of appetite. However bulky food of low digestibility—though available *ad libitum*—can obviously cause restriction of food intake. This may be so in the ruminant in late pregnancy, for example, when the enlarged uterus (and perhaps also abdominal fat deposits) limits the gut capacity. It is doubtful that an animal in this condition should be considered as having appetite unsatisfied. Rather, the hormonal adjustment of tissue uptake to what food is available may be supposed to satisfy the appetite centre—as when food intake is reduced under heat stress.

Two ways in which appetite may be modified are illustrated in Fig. 2.3, which shows seasonal changes in body weight (reflecting appetite) of ferrets on *ad lib.* feeding of a diet of constant composition, and the effect of treatment with progesterone. The seasonal changes are due to photoperiod change, and occur when environmental temperature is held constant. The progesterone effect has also been shown in the pig and the rat, and (though

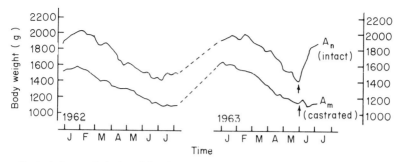

Typical changes in body weight of litter-mate castrated (A_m) and intact (A_n) male ferrets during two seasons and after progesterone implantation. Progesterone implanted at arrow.

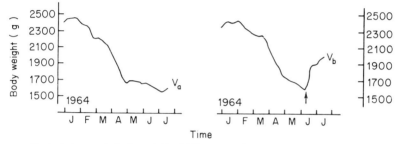

Typical changes in body weight of litter-mate intact male ferrets during one season and after progesterone implantation. Progesterone implanted at arrow.

Fig. 2.3 Body weight changes in entire and castrate male ferrets, fed *ad lib.*, occurring under the influence of natural photoperiod (England), and during treatment with progesterone. (Hammond, J. Jr. and Lawrie, R. A. (1967). *Journal of Endocrinology*, **38**, 243.)

shown here in the male) may be considered a device for producing body stores to meet the needs of late pregnancy and lactation.

This effect of progesterone—at least in the ferret—requires the synergistic action of small amounts of oestrogen. The effect of oestrogen alone (in larger amounts) is to decrease appetite. The action of oestrogen as a growth promoting agent in sheep and cattle (page 106) would seem to be by altering other hormone levels, so that the 'appetite' of fatty tissue for the available food is reduced—to the advantage of muscle growth.

Development

As the fertilized egg evolves, different tissues are formed, these group to form organs, and the plan of the animal is laid down. Subsequently, by differential growth of the parts, the shape of the adult is obtained.

During growth, there is critical timing of different stages of development. Thus, the sex determination of the hypothalamus (p. 17) occurs during a particular period. Again, the young of either sex contains the rudiments of both male and female reproductive organs, further development of which requires stimulation at a critical time. For example, treatment of a female ferret prenatally with male hormone will cause external masculinization and growth of a penis, with an os penis bone, which is lacking in the normal female, and cannot be induced later in life.

True molar teeth are not present in the milk dentition: they are normally erupted only when the jaws have grown large enough to accommodate them. But when R. A. McCance severely restricted the growth of pigs (p. 160) molars were still erupted at the usual age. Puberty results from a decreased sensitivity of the hypothalamus to inhibitory feedback from the gonad. It might seem that this change was not chronologically determined, because the age of puberty is affected by nutrition. However, this may well be because of the extra inhibitory effect of nutritional (and other) stress (see Fig. 4.6).

In the adult animal, cells of some tissues (such as skin or blood cells) die and are replaced at frequent intervals. If part of the liver is removed, liver cells multiply to replace the amount lost. Other cells, however, do not multiply; if fatty tissue is removed it is not replaced.

Growth is due partly to cell multiplication, partly to increase in cell size, and partly to the deposition of extra-cellular connective tissues. Nerve cells and skeletal muscle fibres are all formed early in life. In the early stages of muscle growth the increase is mainly in fibre number (Fig. 2.4); later the increase is entirely in the size of the fibres (see Fig. 4.40). The mouse is born in a rather immature state, before the multiplication of brain cells is

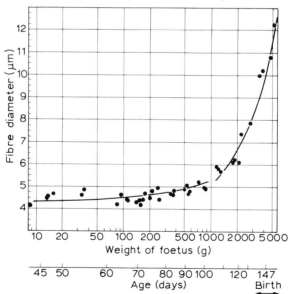

Fig. 2.4 Muscle fibre diameters of foetal sheep. Enlargement of the muscle fibre begins at about the end of the third month of pregnancy; up to that time the muscle cells have been increasing in number. (Joubert, D. M. (1955). *Nature, London*, **175**, 936.)

complete; if it is severely undernourished in late pregnancy and early lactation the number of brain cells is permanently diminished. This is also the case with the secondary wool follicles in the lamb; it is only within a critical period that they can be formed.

A similar example of the importance for development of particular periods in early life is shown by an experiment of McCance (Fig. 2.5): restriction of milk intake of rats in early life permanently alters their capacity for growth—perhaps by altering the sensitivity of some hypo-thalamic feedback.

Probably the number of fat cells, like the number of brain or muscle cells, is determined quite early in life, but this is not easily determined. As a pig, for example, grows and fattens there is an increase in both the number and the average diameter of fat-containing cells (adipocytes) in the layer of subcutaneous fat. The adipocytes are recognizable as contain-ing fat vacuoles; earlier, they are pre-adipocytes. Possibly the number of these is determined in early life, and their transformation later initiated by some timing mechanism.

It should be mentioned that not all fatty tissue behaves in the same way. Some species, particularly hibernators, have brown fat as well as ordinary adipose tissue. Whereas the latter releases free fatty acids for the use of other tissues, brown fat, in response to the 'emergency' stimulus of adrenalin, oxidizes its reserves *in situ*, and so warms the perfusing blood. The new-born lamb—but not the piglet—has brown fat which serves as a defense against cold stress, and subsequently disappears. Again, when general fat reserves are drawn upon in nutritional stress, the fat content of bone increases, as red marrow is replaced by yellow.

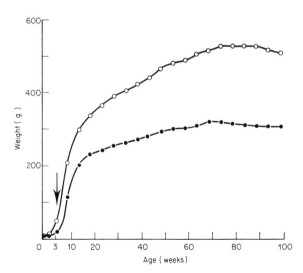

Fig. 2.5 Body weights of rats suckled in small ($\bigcirc$) and large ($\bullet$) litters. $\downarrow$ = weaned. (McCance, R. A. and Widdowson, E. M. (1962). *Proceedings of the Royal Society, B,* **156,** 326.)

The pattern of growth

The increase of body weight follows a characteristic pattern which can be represented in several ways (Fig. 2.6). Plotting weight against time, growth follows an S-shaped curve; it proceeds at an increasing rate with time for a while (until the inflection of the curve is reached) and then slows — to cease at the mature weight. Plotting of the growth increments made in unit time, against time, gives the pattern of Fig. 2.6(b), maximum rate of gain corresponding to the point of inflection of the S-curve. If, however, gain is represented as the fraction by which the body mass adds to itself in unit

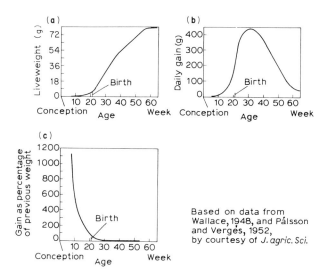

Fig. 2.6   Three types of growth curve for lambs of balanced sexes. **(a)** The actual growth curve. **(b)** The gain-per-unit-of-time curve. **(c)** The percentage increment curve. (Palsson, H. (1955). In *Progress in the Physiology of Farm Animals*. (J. Hammond, Ed.) Butterworth, London.)

time — see Fig. 2.6(c) — growth can be seen to proceed ever more slowly from quite early in prenatal existence.

A possible analogy for this S-shaped curve derives from considering the consequences of size change for an individual cell. The cell exchanges material with its surroundings through its surface — which is large in relation to its mass when the cell is small, and small when the cell is large. Part of the exchange through the surface is by diffusion, part is by active processes whereby the cell preserves differences in concentration between its interior and the exterior. Thus there will be an optimum size for exchange across the surface, analogous to the inflection point of the growth curve.

Economically, the time of inflection of the growth curve is of importance since the amount of food required to maintain the animal increases with its size (though not proportionately). After growth starts to slow, the proportion of food consumed which is used for growth will decrease. However, the composition of the animal, and cost of food, at different stages of growth are also important.

As Fig. 2.7 shows, growth also entails change of proportions. These changes are of relative size of the different parts, and of the various tissues within each part. The changes are brought about by different parts growing at different rates. Some, such as the head—particularly the brain—grow rapidly in early life (Fig. 2.7); later others, such as the limbs, make faster growth and so form a larger part of the whole. Each part and tissue follows an S-shaped curve; the inflection of the curve comes later in late-developing parts.

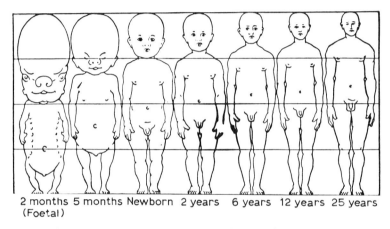

2 months 5 months Newborn 2 years   6 years   12 years 25 years
(Foetal)

Fig. 2.7 Changes in the body proportions of man as he grows up.

It is broadly true that organs vital for the maintenance of life—for example the brain, which co-ordinates body activities, and the gut, upon which the rest of postnatal growth depends—are early developing, and that commercially more valuable parts (muscle, fat, and udder, for instance) develop later. However the late development in evolution of ruminant digestion is reflected in later growth of ruminant stomachs, and some early postnatal deposition of subcutaneous fat may be important for maintenance of body temperature.

Efficiency of growth is sometimes measured by the conversion factor— units of food eaten to produce unit increase of body weight. Because of the nature of the growth process this is not a very good index; in the early stages of growth a larger part of liveweight is made up of gut contents and of offal. Deposition of protein (in muscle, for example) is accompanied by the retention of four parts of water to one of protein, whereas fat—mainly laid down in later stages—is not accompanied by water. However, fat formation is energetically a rather extravagant process, and production of excess fat is certainly economically inefficient.

The growth pattern of pregnancy is shown in Fig. 2.8. The placenta, and particularly the maternal part, is early-developing and the foetus late-

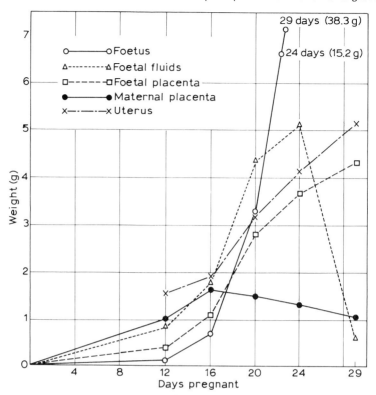

Fig. 2.8   Growth of the uterine contents during pregnancy in the rabbit. (Hammond, J. (1937). *School Science Review*, No. 72, 548.)

developing, so that it outgrows the placenta. Indeed, if the rabbit pregnancy is prolonged beyond day 35, the foetuses die of oxygen deficiency. Thus placental size tends to become a limiting factor as foetal size increases. Concentration of nutrients in the maternal blood and rates of foetal and maternal blood supplies to the placenta, besides the area for diffusion between them, may restrict availability of nourishment to the young. With Shire × Shetland crosses (Chapter 3) the large foal had a much larger placenta; with strains of mice, McCarthy (1965) has shown that increased foetal size in hybrid young is generally accompanied by increased placental size.

In the rabbit, litter size is variable, and gestation is generally longer with smaller litters. After longer gestation the young are naturally larger, and more mature in their body proportions—that is to say, the head, an early developing part, forms a smaller proportion of the whole body (Fig. 2.9). But, at the same gestation length, body size and proportions are affected by litter size. In a large litter the young are smaller, and their heads form a larger proportion of the whole. Though chronologically of the same age, anatomically the large litter are 'younger'. That young of the large litter

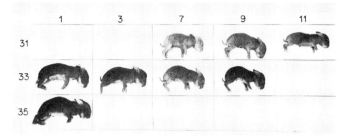

Fig. 2.9 Newborn rabbits, showing how the number in the litter (*above*) affects the weight and body development. The duration of pregnancy in days (*side*) decreases with the number in the litter, but this has only a slight effect on the size of the young. (Wishart, J. and Hammond, J. (1933). *Journal of Agricultural Science*, **23**, 463.)

are smaller is to be explained by the fact that more young are competing for a limited amount of nutrition supplied by the mother. That body proportions are different is to be explained by unequal competition between the various tissues. The same situation applies postnatally when young are solely dependent on the mother's milk (Fig. 2.10), and might be expected to apply to adult body proportions if nutritional level is sufficiently low and long sustained.

In general, each part of the body, and each tissue composing it, follows an S-shaped curve—the peak growth rate coming at different times for

Fig. 2.10 The effect of the amount of milk supply on growth. (*Below*) Young rabbits of an inbred strain at one month old which had access to the whole milk supply of a doe, (*above right*) shared it with two others, and (*above left*) shared it with four others.

different parts, and depending on the plane of nutrition. Figure 2.11 is a general summary of the growth process. Furthermore, the spatial distribution of these peaks of growth is not random; a wave of growth spreads backward along the trunk, and secondary waves, starting low in the limbs, spread upward to meet it in the loin—which develops late. Figure 2.12 shows how the growth curve inflection comes later in the bones higher up the limbs, and Fig. 6.14 (p. 164) that bone is earlier developing than muscle, and muscle earlier than fat.

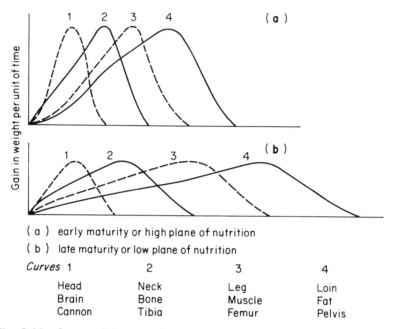

(a)   early maturity or high plane of nutrition
(b)   late maturity or low plane of nutrition

Curves 1	2	3	4
Head	Neck	Leg	Loin
Brain	Bone	Muscle	Fat
Cannon	Tibia	Femur	Pelvis

Fig. 2.11 Curves of the rate of increase in weight which show the order of development of the different parts and tissues of the body. The way in which the changes in shape and body proportions are affected by early and late maturity and by the level of nutrition are also shown. (Palsson, H. (1955). In *Progress in the Physiology of Farm Animals*. (J. Hammond, Ed.) Butterworth, London.)

Suboptimal nutrition clearly affects more the growth of later than earlier-developing parts of the body; it seems natural to extend the generalization to tissues. Clearly it applies to the relative development of bone and fat; but between bone and muscle development there seems to be a close relationship—as there is in their functioning. The extent of mineral deposition in bone depends upon the load placed upon it, so the weight of bone must be related to the weight of animal. The length growth of the muscle is obviously determined by the length growth of the bone over which it works, and the muscle's size by the work that it is called upon to do by its nerve supply; a muscle wastes if its nerve is cut, and experimentally, its character can be altered by altering its nerve supply. Nevertheless,

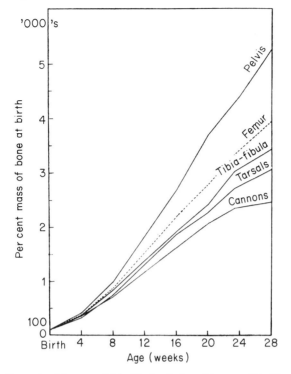

Fig. 2.12 Growth of bones of hindlimb of pig with age. (McMeekan, C. P. (1940). *Journal of Agricultural Science*, **30**, 276.)

in the adult there is not a close relationship between bone and muscle weight; and in the young animal under extreme conditions bone can grow while muscle wastes (Pomeroy, 1941).

Statisticians studying growth like, if possible, to work with straight-line relationships. Such a relationship was found by Huxley to give a good description of the changes in proportion, for example, of a crustacean as it went through a series of moults. If the logarithm of the weight (or length) of one part, X (say a segment of a limb), was plotted against the logarithm of the length—or weight—of another part, Y (a different segment of that limb, perhaps), then the points fell on a straight line. Mathematically, that statement can convert into the equation $Y = aX^k$, in which a and k are constants. In this equation the value of a depends upon the units of measurements used, but k is very useful as a measure of the relative growth of the two parts X and Y; if $k = 1$ there is no change of proportion with growth, and the greater k is, the greater (i.e., the later-developing) the growth of Y relative to X (and conversely if k is less than one).

The equation in fact represents the relationship between two quantities growing steadily at different rates of compound interest. However *time* (growth *rate*) does not feature in the equation, and it can easily be shown that, if the peak of growth for part Y comes later than that of part X, then the value of k must be greater than one.

Clearly the equation does not describe the situation in Fig. 2.8 in which growth of one part (the maternal placenta) has ceased while that of another (the foetus) still proceeds rapidly. Nor can it be mathematically exact where one of the two parts compared itself consists of two or more portions with different relative growth rates. However, it can be a useful approxi-

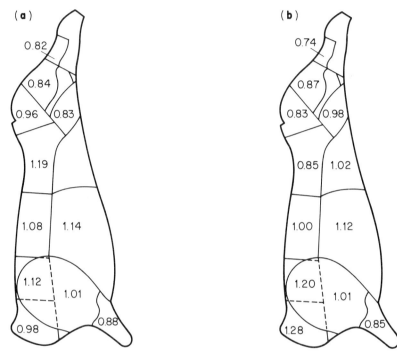

Fig. 2.13 Growth coefficients (k values) for **(a)** bone and **(b)** muscle in commercial joints from carcases of young bulls. (Adapted from Berg, R. T., Anderson, B. B. and Liboriussen, T. (1978). *Animal Production*, **27**, 51, 71.)

mation, and Fig. 2.13 illustrates, in more detail, and in another way, the pattern of growth shown in Fig. 2.12 and described in the paragraph referring to it (p. 35). In calculating the values of k shown in Fig. 2.13, the weight of bone in a joint of the carcase has been compared with that of the total weight of bone in the carcase, and of muscle in a joint with that of total carcase muscle. It will be noted that the growth pattern for muscle and bone are found to be not quite the same.

Growth and sex

The change in body proportions which occurs with increase of size is not always due to steady differential growth. In the case of a prawn, analysed by Huxley, plotting logarithmically length of claw against size of body, male and female claw grew alike up to a certain size; above that size the claw of the female continued at the same relative growth rate, the claw of the male grew more rapidly—its relative growth rate had changed.

Similarly there is a change in relative growth of avian and mammalian organs dependent on sex hormones. These start as early-developing parts, and so continue in the castrate, becoming relatively smaller as the animal grows; but in the entire animal, as puberty approaches, their relative growth rate is increased; they become late-developing.

Generally there is a sex difference in body size; the male is usually heavier than the female (in rabbits and hamsters the reverse is true). The greater size is attained both by faster growth and by growth continuing for longer. Puberty appears at about the inflection of the growth curve; so in general it occurs earlier in females than in males, and earlier in breeds of small adult size than in larger ones. Pregnancy occurring well before body growth is complete can divert nutrients that would otherwise be available for growth, and so stunt adult body size.

As well as in size, there is sex difference in body conformation and degree of fatness. In general the female is nearer to the neuter, or castrate; but the castrate of either sex is fatter, and has usually greater length of bone. The greater muscular development of the male—a desirable feature from the viewpoint of meat production—is presumably to be attributed to the 'anabolic' action of male hormone. It is to some extent counterbalanced by the features of male conformation—a heaviness in the neck and shoulder—which are not the most valuable parts of the carcase. However this is a late developing character (see Fig. 8.4—how it is depressed by under-nutrition, giving the bull a very feminine appearance).

Reasons for the difference in body size are probably rather complex. The experiment of Zawadowsky in chickens (see Fig. 7.13) would appear to show that genetic composition rather than sex hormone secretion was responsible. On the other hand, in the ferret the male is twice the size of the female, but castration at birth prevents this size difference developing. In the rat, Perry and others found no growth promotion by treating females with testosterone after weaning, though removal of the ovary at that time did enhance growth rate. However a single injection of testosterone soon after birth—which masculinized the hypothalamus—with subsequent removal of the ovaries, resulted in a growth rate very like that of the male.

The existence of these sex differences in growth can lead to some confusion about 'maturity'. From the butcher's point of view a beast is mature when there is sufficient muscle on the bone, and fat on the muscle—and is overmature when there is too much fat. Fat is a late-developing tissue, and the castrate is fatter than the female: so one might expect the heifer to 'mature' later than the steer. However the steer is bigger and goes on growing longer; so that, in well fed animals at equal weight, the proportion of fat in the carcase becomes too great at lower weight in the heifer—a fact witnessed by the prices paid (see Fig. 4.43).

References

HUXLEY, J. S. (1932). *Problems of Relative Growth.* Methuen, London.

MCCARTHY, J. D. (1965). Genetic and environmental control of foetal and placental growth in the mouse. *Animal Production,* **7**, 347.

PERRY, B. N., MCCRACKEN, A., FURR, B. J. A. and MACFIE, H. J. H. (1979). Separate roles of androgen and oestrogen in the manipulation of growth and efficiency of food utilization in female rats. *Journal of Endocrinology,* **81**, 35.

POMEROY, R. W. (1941). The effect of submaintenance diet on the composition of the pig. *Journal of Agricultural Science,* **31**, 50.

Further Reading

COLE, D. J. A. and LAWRIE, R. A. (Eds) (1975). *Meat.* Butterworth, London.

LODGE, G. A. and LAMMING, G. E. (Eds) (1968). *Growth and Development of Mammals.* Butterworth, London.

MCCANCE, R. A. and WIDDOWSON, E. M. (1974). The determinants of growth and form. *Proceedings of the Royal Society of London, B,* **185**, 1.

3   Horses

The breeding season

The horse is an animal in which the natural breeding season occurs during the spring and summer months, i.e. during the period of increasing hours of daylight. When mares are taken from the northern to the southern

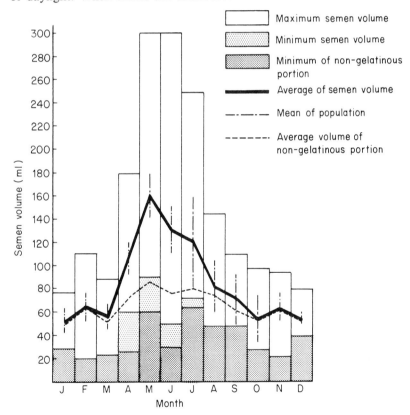

Fig. 3.1   Seasonal variation in volume of stallion semen. The variation is due mainly to variation in the gelatinous seminal fluid. (Nishikawa, Y. (1959). *Studies on Reproduction in Horses*. Japan Racing Association, Tokyo.)

hemisphere they change the time of their breeding season to fit the seasons of the new surroundings. Wild and semi-wild breeds have a distinct breeding season when the days are increasing to maximum length, so that their young are born in the spring of the following year. With domestication, the duration of the true breeding season increases but even with modern breeds the anoestrous or non-breeding season is not entirely eliminated, although some individual mares will breed the whole year round. For example, about half of the Welsh and Shetland pony mares investigated at Cambridge would breed throughout the whole year while the other half failed to breed from October to March.

The main factor affecting the onset of breeding activity in mares, and the volume of semen produced by stallions, is an increase in hours of daylight. This is reflected in periods of maximum breeding activity related to latitude. For example, the maximum percentage of services producing foals occurs in May to July in Canada (57°N), in April to July in the U.S.A. (40°N) and in both April–May and October–November in India where the sun passes overhead twice a year. South of the equator the situation is reversed, so that the maximum number of fertile services occurs in November and December in Australia and New Zealand (30°–40°S). In stallions the major factor in the increase in volume of semen in the breeding season is an increase in the gelatinous seminal fluid (Fig. 3.1).

The oestrous cycle

The occurrence and duration of heat can be tested by running the mares with a vasectomized stallion (in which the ducts from the testes to the penis have been cut).

The length of the heat period averages about 7 days but may vary from 3 to about 30 days. In a cold dry spring, and especially in very young or very old mares in poor condition, the heats may last a long time, 10 to 15 days or more being not uncommon. This is due to the slow ripening of the follicle under these conditions. As the breeding season advances, however, the heats tend to get shorter, so that by May–July (Northern Hemisphere) they last only 5 or 6 days on the average (Fig. 3.2). Under the conditions which give rise to long heat periods there is lowered chance of fertility (see page 44).

The time between the beginning of one heat period and the next is commonly stated as 3 weeks but this is only true when the mare has an average heat period of 5 days. If the heat period is longer than this then the interval between the beginning of one heat and the beginning of the next will be correspondingly longer. The best way to calculate the expected onset of the next heat is by taking the interval as extending 16 days after the end of the last heat. Unfortunately even this period—the interoestrous or dioestrous period—is also variable, but is usually within the range of 14 to 19 days.

After foaling, mares usually have a 'foal heat'. Most come in heat 4 to 7 days after foaling (range 4 to 17 days) and this has led to the usual

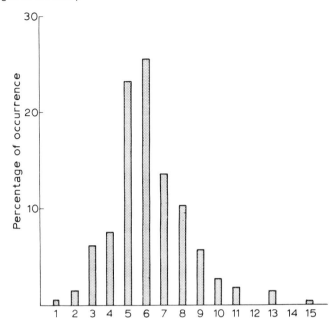

Fig. 3.2   Distribution of duration of heat in 283 oestrous periods in 25 mares during the period May to July. (Nishikawa, Y. (1959). *Studies on Reproduction in Horses*. Japan Racing Association, Tokyo.)

practice of trial and service on the 9th day. Fertility at this heat period has been reported as being lower than at the next one, and the incidence of subsequent abortion higher. However, delaying mating by about 15 days, at the heat period which followed regression of the corpus luteum induced by prostaglandin treatment (p. 15), seems not to improve fertility (Burns *et al.*, 1979).

Changes in the ovary

The structure of the ovary of the mare is peculiar. In early life it is like that of most other species, with a large surface on which ovulation can occur. As it develops it bends inwards and becomes bean-shaped and covered with a thick serous coat except in a limited space in the centre of the bend— the ovulation fossa. Follicles, within which eggs are matured, may develop anywhere in the ovary, but as the heat period progresses they expand towards the ovulation fossa, and ovulation occurs only at the fossa.

As the breeding season approaches several follicles enlarge in the ovary; one of these will continue to enlarge while the others regress. As this larger follicle increases further in size it produces sufficient hormone to induce oestrous behaviour. At the begining of the season such a follicle (and oestrous behaviour) may persist for some time; eventually the follicle either ovulates, or regresses and is replaced by another. Later in the season the follicle continues to grow towards the ovulation fossa and by the 5th day it is quite large. It then ruptures and the egg is washed out in the follicular fluid and enters the funnel of the oviduct or Fallopian tube. The mare goes out of heat about 24 hours later. If a fertile mating has occurred, the egg is fertilized in the upper third of the tube within a few hours of being shed. On the 6th and last day of heat (in the example shown in Fig. 3.3) the cavity from which the egg was shed is seen. In this cavity there accumulates blood from the vessels broken at the time of ovulation. By 2 days after the end of heat the cells of the walls of the follicle grow in a fern-like formation, until by 8 days they form a solid yellowish body, the corpus luteum or yellow body. The changes up to 4 or 5 days can be determined by palpating the ovary with the fingers through the wall of the rectum. After 5 days the corpus luteum is no longer palpable but the ovary containing it is usually about twice the size of the other.

The developing foetal membranes act on the uterus and prevent it from causing early regression of the corpus luteum. If there is no foetus the corpus luteum rapidly decreases in size after about 16 days, and the fall in level of progesterone allows the pituitary to stimulate development of another ripe follicle. This atrophy of the corpus luteum becomes apparent 2 days before heat and is obvious by the first day of heat.

The exact time at which the egg is shed is important because the sperm of the male live for but a short time in the female tract. Hence the chances of fertility are increased if mating occurs shortly before the egg is shed. If the follicle is ruptured by pressure, in some mares on the 2nd day and in others on the 4th day of heat, the heat lasts just under or just over 24 hours, which confirms the observation that the shedding of the egg occurs about 24 hours before the end of heat, regardless of its length. In Japan, Nishikawa has studied the time of normal ovulation relative to the end of heat. Sixty-five per cent of mares ovulated 1 day before the end of heat and 25 per cent 2 days before.

Fertility and sterility

It is necessary that the mare should be mated just before the egg is shed if optimum results are to be obtained. If the mare is mated only on the first day of a long heat period of 10 days the sperm will have perished before the egg is shed on the 9th day. With Thoroughbreds, where the mares are served 2 or 3 times on alternate days in each heat period, the fertility is high (66 per cent). The pony stallion running out on the hills with mares all the time has a still higher fertility (95 per cent). In mares with long heat periods it has been found an advantage to inject intravenously a hormone preparation rich in luteinizing activity. The hormone commonly used is prepared from the urine of pregnant women and treatment with 500 to 2000 i.u. of HCG causes ovulation in about 30 to 50 hours, and so saves repeated services.

It has been found with most species of animals that the egg does not remain capable of fertilization for more than a few hours after it has been shed. If the mating is made after the egg has been shed, as would occur

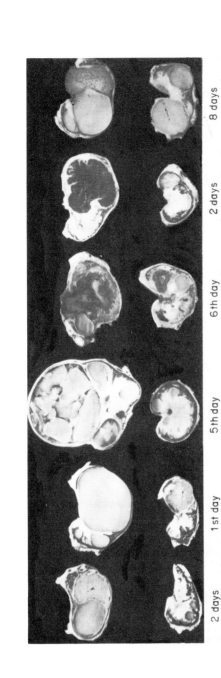

Fig. 3.3 Ovaries of mares at different stages of the oestrous cycle. The two ovaries (right and left) of an animal are shown one above the other. The top row shows the ripening follicle which enlarges greatly during the heat period and ruptures about one day before the heat ends. From the walls of the ruptured follicle a corpus luteum is formed; this is at first dark from blood which breaks into it, and then becomes paler as the luteal tissue develops. (Hammond, J. (1938). *The Sechenov Journal of Physiology of the U.S.S.R.,* **21**, 193.)

right at the end of heat in the mare, there is a reduced chance of the sperm reaching the egg before it becomes incapable of fertilization. Table 3.1 shows the results of two experiments, one conducted in Cambridge and one in Japan (cited by Nishikawa and Hafez, 1962), in which mares were given a single mating with a fertile stallion on known days in the heat period or relative to the time of ovulation.

It will be seen that the chances of fertility are highest when mating is made from 0 to 3 days before ovulation or 1 to 4 days before the end of the heat. Earlier or later matings result in lower fertility.

It has been estimated that the stallion produces 4 to 6 thousand million spermatozoa per day, sufficient to permit use of an average stallion once or twice a day during the breeding season. All the matings in the first experiment were made by one stallion. Individual stallions, however, vary

Table 3.1 Fertility of mares mated at different times.

1. Cambridge								
Days from end of heat	−13 to −9	−7	−6	−5 to −3	−2		−	+1
Number of mares mated	4	7	6	11	9	Time when egg is shed	5	0
Per cent fertile	0	29	50	64	67		20	−
2. Japan								
Days from ovulation	−11 to −7	−5	−4	−3 to −1	0		+1	+3
Number of mares inseminated	0	30	20	377	256		13	1
Per cent fertile	−	10	40	60	60		54	0

much in the number and activity of the sperm in their semen. In those with relatively few, weakly motile, sperm in the ejaculate the chances of successful fertilization are reduced below those for stallions with large numbers of highly motile sperm, particularly if mated early relative to the time of ovulation. In late oestrus the muscular tone of the cervix (entrance to the uterus) changes and it becomes extremely sensitive to touch, so that distinct grasping contractions can be felt if a hand is inserted. These contractions probably occur during mating. Some mares may strain after service, and evacuate most of the semen, so it is a common practice to walk a mare directly away after service and to keep her walking for a while. While walking she cannot strain.

Sterility due to infection of the genital tract is liable to occur in mares, especially those of the light breeds, through injury to the vulva which allows them to suck air into the uterus. By stitching the upper commissure of the vulva, the normal semi-anaerobic condition of the tract is restored and the infection is cleared up.

The stallion and artificial insemination

While the fertility of any particular stallion may vary slightly from season to season, there can be no doubt that individual differences exist in fertility. They are due to differences in the number and activity of the sperm contained in the semen. Although only one sperm is required to fertilize the egg, it appears (by analogy with other species) that a large number, possibly of the order of 3000 or so, are required in the oviduct soon after ovulation in order to provide a good chance of fertilization.

It is now an easy matter to test the quality of the semen of a stallion before the breeding season begins. An artificial vagina (see page 73), is held against the flank of the mare so that the stallion serves into it and the whole of the semen can be collected in an uncontaminated state. From such collections it has been determined that the normal stallion ejaculates about 50 to 100 ml of fluid containing about 5000 to 15 000 million sperm. If frequent collections are made, the volume and sperm density both fall rapidly. It is generally considered that a dose of 2500 million sperm in 50 ml is adequate for insemination; but (see below) this allowance may be unnecessarily generous.

There are difficulties about the storage of stallion semen; also there are strong breed society prejudices in the British Commonwealth and in the United States against the use of artificial insemination. Consequently the technique has been little used or developed. Successful storage of stallion semen by freezing was first achieved in Japan by Nagase and others. They concentrated the semen by centrifugation, equilibrated it in a medium containing glycerol, and then froze it by the pellet method (p. 75); the thawed semen was then diluted before insemination. More recently, Martin *et al.* (1979) used a method developed for the pig and obtained pregnancy in 12 of 19 mares inseminated. The sperm-rich fraction of the ejaculate was centrifuged, then appropriately diluted and frozen in enlarged (4–5 ml) Cassou straws (see Chapter 4): a single straw, containing about 200 million motile spermatozoa, was used for each insemination.

Diagnosis of pregnancy

The foetal membranes in contact with the uterine lining prevent the regression of the corpus luteum that would otherwise occur about 16 days after ovulation. A really firm placental attachment between mother and foal is only formed after about 50 days of pregnancy; but long before this the foetal chorionic cells which secrete the mare serum gonadotrophin migrate into the lining of the uterus. So it comes about that a hormone of foetal origin, and with such a large molecular size that it does not cross the placenta, is found in the blood of the dam.

From the 40th day of pregnancy additional, accessory, corpora lutea are formed; but by the end of the fifth month all corpora lutea regress and the placenta takes over the role of production of progesterone until the end of pregnancy.

The MSG is generally believed to stimulate the formation of these accessory corpora lutea; but the follicle growth involved is not accompanied by much oestrogen production. MSG appears in appreciable quantities in the blood by about the 40th day of pregnancy, reaches a maximum at about day 70 and then falls rapidly (Fig. 3.4). Oestrogen excretion in the urine, however, only begins to rise above the level found at oestrus from about day 90. This oestrogen does not derive from the ovaries, which appear quite inactive in the second half of pregnancy until towards its end, when follicles enlarge in preparation for the 'foal heat' (Fig. 3.5).

The cells which produce the MSG are 'foreign' to the mother and provoke an immune reaction in her; this probably explains the variation in amount of the hormone (Fig. 3.4) and why more should be present in the

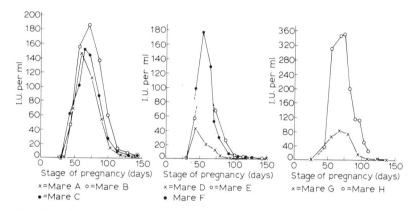

Fig. 3.4 Pregnant mare serum gonadotrophin (MSG) in blood of mares during pregnancy. (Day, F. T. and Rowlands, I. W. (1940). *Journal of Endocrinology*, **2**, 255.)

first pregnancy (before any immunity has developed). It may also explain why they are much lower in mares in foal to a jackass (Clegg *et al.*, 1962).

If a mare fails to come into season three weeks after service, it might be assumed that she is safely in foal, because persistence of the corpus luteum had prevented the ripening of follicles and a return to heat. However, if she had been served early in a long heat period she would have been unlikely to conceive, and the corpus luteum eventually formed would not have died away by three weeks after the date of service. This explains why many mares are missed. In addition, some mares have only a few heat periods during the season and may not show any further heats even though they are not in foal. Hence some reliable method is required for the early diagnosis of pregnancy. There are four main ways in which this can be done:

(1) *Palpation of the uterus through the wall of the rectum* By this method an experienced person can detect pregnancy with a fair degree of certainty as early as the 20th day. In a pregnant mare at this stage the uterine horns are turgid, and not flaccid as during the cycle, while the

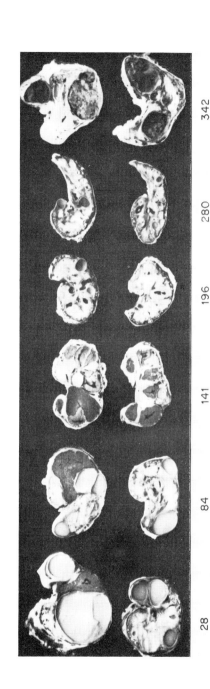

28   84   141   196   280   342

Days pregnant

Fig. 3.5 Ovaries of mares at different stages of pregnancy. In the early stages there is only one corpus luteum, from which the fertilized egg came; by 141 days, however, there are a large number of corpora lutea produced by luteinization of all the small follicles without rupture; at 196 and 280 days no follicles or corpora lutea are present, but at 342 days follicles are being formed in preparation for the heat which occurs shortly after parturition. (Hammond, J. (1938). *The Sechenov Journal of Physiology of the U.S.S.R.*, **21**, 193.)

embryo can be felt in one horn just above the cervix as a swelling about the size of a bantam's egg. By the 45th day it is the size of a goose's egg and by the 70th day the size of an ostrich's egg. With experience one can tell the stage of pregnancy within a week or so by this means (Day, 1940). This is particularly useful for selecting hill pony mares for bleedings for pregnant mare serum hormone, which rises to a peak yield about the 70th day of pregnancy.

(2) *The blood progesterone test*   It is feasible to measure the concentration of progesterone in blood plasma (p. 14). If a blood sample is taken 17 to 22 days after ovulation and the concentration of progesterone in it is less than $1.5 \, \text{ng ml}^{-1}$ (a nanogram is one millionth of a milligram) it may safely be assumed that the corpus luteum has regressed and that the mare is not pregnant. A higher level indicates an active corpus luteum; but about one in four of mares indicated by this test to be pregnant will not foal, owing (probably) to early embryonic mortality (Palmer, Thimonier and Lemon, 1974).

(3) *The serum gonadotrophin test*   MSG hormone is present in the blood in appreciable quantities between the 40th and 100th day of pregnancy; its presence is detected by injecting a small volume of serum from the mare into immature mice—when the MSG, if present, will stimulate the ovaries of the mice, and enlarge their uteri, within a few days. A false positive could be obtained from this test, since abortion of the foetus will not necessarily entail loss of the cells producing the hormone, which in turn may be sustaining luteal tissue in the ovary. In such a situation treatment with prostaglandin should ensure a return to oestrous.

(4) *The mucin test*   One disadvantage of the two previous tests is that they cannot be done in the field. The mucin test, however, can be made on the farm and the results known immediately. Although the test can be made from about 1 month after conception, its reliability is enhanced after the third month as the criteria used are more clearly expressed. However the test should not be made late in autumn or in winter as those mares which do not breed during the winter months may show a similar pattern.

The basis of the mucin test is the change which occurs in the mucus produced from the cervix. At the time of mating the mare is under the influence of oestrogen, which causes the surface of the vagina to be covered by a thin watery slime or mucus, and the cervix is open. Following mating and ovulation, a corpus luteum forms and produces progesterone. This hormone causes the cervix to close and the mucus to thicken, so sealing up the cervix and preventing bacteria from entering the uterus. These processes are intensified as pregnancy advances.

The test for pregnancy can be made in either of two ways. First, a speculum is inserted in the vagina and a sample of the secretions in the cervix is obtained by a long-handled paint brush. This is then painted on a glass slide and stained with methylene blue. In pregnancy the mucus is adhesive and contains globules, and the smear may contain cells with cilia or hair-like lashes. Alternatively a hand may be inserted into the vagina and the fingers drawn lightly over the upper end of the vagina and cervix. If the surface feels dry and sticky the mare is probably in foal.

The duration of pregnancy

Although the duration of pregnancy is usually reckoned as 336 days, there is considerable variation from this average. Factors which are known to affect the duration are the month in which the foal is due to be born, the genetic make-up or genotype of the foetus, and the number born.

Wellman in Hungary and Gonnermann in Germany report durations of pregnancy which increase from 320 days for foals born in October to 340–341 days for those born in May, and decrease to 323 days for those born in July. Figure 3.6 shows a marked seasonal variation in Welsh pony mares

Month in which due (336 days) to foal	Duration of the pregnancy (days)								
	Average−			Average	Average +				
	11 to 15	6 to 10	1 to 5	336	1 to 5	6 to 10	11 to 15	16 to 20	21 to 25
March 16–31								●	●
April 1–15									
April 16–30							●	●	
May 1–15				●	●	●●			
May 16–31									
June 1–15			●		●				
June 16–30	●	●	●						

Fig. 3.6 Variation in the duration of pregnancy in Welsh pony mares due to foal at different times of the year. (Hammond, J. (1938). *The Sechenov Journal of Physiology of the U.S.S.R.*, **21**, 193.)

in Great Britain. The pattern differed from that in Hungary, although it occurred regularly every year.

The reasons for these differences are obscure; two have been suggested. One is that different seasonal feeding conditions affect the time of foaling and the other that photoperiod acting through the pituitary gland affects the hormone production of the ovaries, and so influences the time of birth. Which, if either, of these is the true explanation must await further investigation, but in the meantime the determination of what does occur may assist the horse breeder in more accurately fixing dates to prepare for mares foaling.

It is well known that the duration of pregnancy of a mare in foal to an ass is some 15 days longer than if she is mated to a stallion. Hence in this extreme hybrid form the foetal genotype can affect the duration of pregnancy. But this can also apply to purebreeds. In 1951, Rollins and Howell published an analysis of the length of pregnancy of 186 pure-bred Arabian horses and showed an important effect of the genotype of the foetus on the duration of pregnancy. The mechanism of this effect of the foetus is not known. Another factor to be remembered is the size of the foetus. In general, heavy foetuses are born earlier than are light foetuses.

Finally, although twin ovulations are not rare, twins in mares are uncommon and are seldom viable; those which are successfully carried are usually born early, before full term.

Growth of the foal

The amount of growth made by the foal is small during the early stages of pregnancy, but from the middle of pregnancy onwards it increases at an ever-increasing rate. It is, therefore, during the latter part of pregnancy that the feeding and management of the in-foal mare is important. At this time the demands of the foetus for minerals—particularly calcium and phosphorus—and for protein and the essential vitamins are at their greatest. Further, the udder tissue of the dam is developing. Hence a well balanced ration adequate to meet these requirements and to keep the dam in a thriving but not over-fat condition is required.

The size of the foal is probably controlled more by the size of the dam than by her nutrition. For example, in reciprocal crosses between the large Shire horse and the small Shetland pony the size of the foal follows that of the dam (Fig. 3.7). The size of the uterus of the small Shetland dam limits the size of the placenta and hence the nutrition and size of the foal to such as she can deliver. The cross-bred foal out of the Shire mare is three times as large at birth as is the cross-bred out of the Shetland mare. At four years of age these differences are still marked, the former still being one and a half times heavier than the latter. The differences are maintained until adult life (14 years old).

These maternal effects on size of the offspring in reciprocal crosses between large and small breeds of horses have been confirmed by Flade in Germany. The same size differences occur in crosses between the horse and donkey; the mule with the large horse as dam is much larger than the hinny with the small donkey as dam. From this it would appear that if size is required in horses one must get a good growth of frame in the foal when it is young. This can best be done by breeding from large-framed mares and seeing that the foal has a good provision of milk in the early stages of life, for milk is one of the best foods to support growth. After weaning, the provision of a balanced ration containing adequate protein of high biological value is essential. A protein of high biological value is one, like that of milk, which contains a balanced proportion of all the essential amino acids. Fish meal is a good substitute for milk at this stage because it contains not only protein of high quality but also a balanced proportion of available minerals. Provision of adequate vitamins, notably vitamins A and D, is also essential.

Not only is the actual growth in size important, but the change in body proportions as the foal grows up is also of interest, for it is in this respect

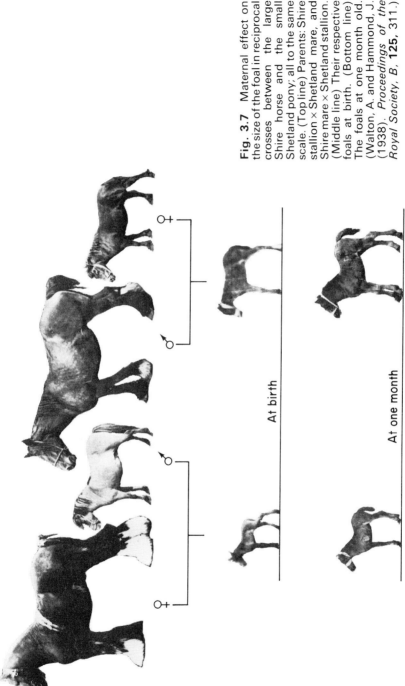

Fig. 3.7 Maternal effect on the size of the foal in reciprocal crosses between the large Shire horse and the small Shetland pony; all to the same scale. (Top line) Parents: Shire stallion × Shetland mare, and Shire mare × Shetland stallion. (Middle line) Their respective foals at birth. (Bottom line) The foals at one month old. (Walton, A. and Hammond, J. (1938). *Proceedings of the Royal Society, B,* **125**, 311.)

that breed improvement for draught, racing or riding purposes has been achieved over many years. Long ages ago the horse was a small marsh-living animal about the size of a dog. Its natural evolution consisted of its taking to drier ground and of the development of speed by increasing the length of leg in proportion to the length of the body (Fig. 3.8). This is seen when the proportions of the skeletons of fossil horses during evolution are compared to a common cranium size. The cranium (eye-ear length) is a part of the body which develops early in life and so it is convenient to use

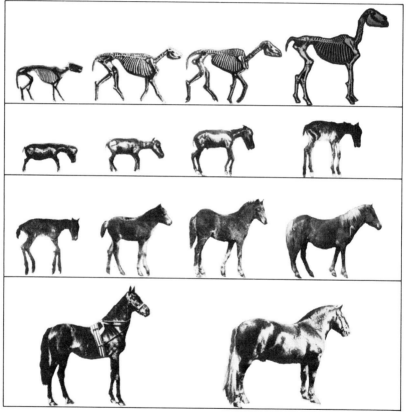

Fig. 3.8 Changes in the proportions of the horse in development and during evolution. In order to show the changes in proportions, all the photographs have been reproduced to the same cranium (eye-ear) length. The changes in proportions during embryonic life parallel those which have taken place during evolution. Reading from left to right. (*Top line*) Evolution: Eohippus; Mesohippus; Merychippus; Equus (Arab). (*Second line*) Development (Welsh pony): 3 months; 5 months; 7 months; 10 months. (*Third line*) Development (Welsh pony): 11 months; 2 weeks after birth; 9 weeks; adult. (*Bottom line*) Evolution: light horse (Thoroughbred-St Simon); heavy horse (Suffolk-Wedgewood). (Hammond, J. (1934). *Proceedings of the 16th International Congress of Agriculture, Budapest*, Section VI.)

as a standard to show how the proportions change. The foetus of the horse today repeats in its development the changes of evolutionary form undergone by its ancestors in the past. The 3-month foetus of the Welsh pony has very much the body proportions of the primitive Eohippus, and the Welsh foetus at 7 months is like that of the fossil Merychippus. Thus, both in evolution during time and in the development of the embryo in the individual today, there is an increase in the proportion of the length of leg, fitting the horse for speed. The changes in proportions which occur after birth in the Welsh pony differ from the previous ones in that the body becomes thicker and deeper and the head and legs smaller in proportion.

Breed improvement in horses today (see Chapter 12) is proceeding along two main channels, namely for speed and for endurance. Selection for speed continues along the main evolutionary channel seen in the development of the foetus. The body proportions of the Thoroughbred are an extension of these changes. Selection for endurance is exemplified by breeds such as the American quarter horse. Here the evolutionary trend is not exaggerated as in the Thoroughbred. Endurance, associated with somewhat shorter and thicker bones and consequent deeper muscles, coupled with certain characteristics of gait and form are sought. A third channel of selection, namely for draught, has virtually disappeared in countries such as Great Britain, Australia and the United States, and good examples are becoming rare. It persists in some countries such as France but not for draught purposes. With mechanization of farms, those areas that traditionally used oxen for draught have persisted with the rearing of cattle for beef production. Those areas which traditionally used horses for draught have persisted with horses for meat production.

References

BURNS, S. J., IRVINE, C. H. G. and AMOS, M. S. (1979). Fertility of prostaglandin-induced oestrus compared to normal post-partum oestrus. *Journal of Reproduction and Fertility, Supplement*, **27**, 245.

CLEGG, M. T., COLE, H. H., HOWARD, C. B. and PIGON, H. (1962). The influence of foetal genotype on equine gonadotrophin secretion. *Journal of Endocrinology*, **25**, 245.

DAY, F. T. (1940). Clinical and experimental observations on reproduction in the mare. *Journal of Agricultural Science*, **30**, 244.

MARTIN, J. C., KLUG, E. and GÜNZELL, A-R. (1979). Centrifugation of stallion semen and its storage in large volume straws. *Journal of Reproduction and Fertility, Supplement*, **27**, 47.

NISHIKAWA, Y. and HAFEZ, E. S. E. (1962). The reproduction of horses. In *Reproduction of Farm Animals* (E. S. E. Hafez, Ed.), chapter 16, p. 266. Lea & Febiger, Philadelphia.

PALMER, E., THIMONIER, J. and LEMON, M. (1974). Early pregnancy diagnosis in the mare by estimation of the level of pregesterone in peripheral blood. *Livestock Production Science*, **1**, 197.

Further reading

EVANS, J. W., BORTON, A., HINTZ, H. F. and VAN VLECK, L. D. (1976). *The Horse*. W. H. Freeman & Co., San Francisco.

NISHIKAWA, Y. (1959). *Studies on Reproduction in Horses*. Japan Racing Association, Tokyo.

ROWLANDS, I. W., ALLEN, W. R. and ROSSDALE, P. D. (Eds) (1975). Equine Reproduction. *Journal of Reproduction and Fertility, Supplement*, **23**.

ROWLANDS, I. W. and ALLEN, W. R. (Eds) (1979). Equine Reproduction. II. *Journal of Reproduction and Fertility, Supplement*, **27**.

ROWLANDS, I. W. and WEIR, B. J. (Eds) (1982). Equine Reproduction. III. *Journal of Reproduction and Fertility. Supplement*, **32**.

4 Cattle and Buffalo

The breeding season

Unlike the mare and the ewe, the cow does not have a clearly defined anoestrous, or non-breeding, season. However, during the winter months breeding activity is lower than in the summer. 'Still' or 'silent' heats (ovulations without oestrous behaviour) occur, as do short heat periods of 6 hours or so. In practice it is found that in cows which calve during the autumn months there is more delay in getting them in calf again than in those which calve in the spring. As a consequence, unless care is taken, a

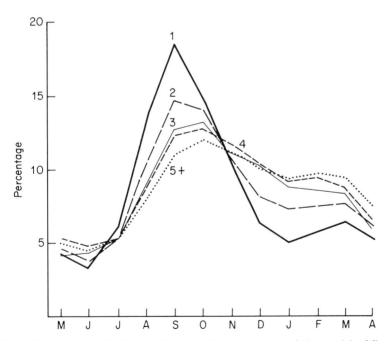

Fig. 4.1 Seasonal distribution (as monthly percentage of the totals) of first, second, third, fourth and later calvings of all Friesians recorded by the English Milk Marketing Board in 1976–77. (Milk Marketing Board. *Report of the Breeding and Production Organization*, No. 28, 1976–77.)

high proportion of the cows soon come to calve in the spring, which is what occurs in nature. To make up for this it is the common practice in dairy herds to mate all heifers so that they calve in the autumn, so ensuring a good supply of winter milk (Fig. 4.1).

It is not easy to separate the influence of photoperiod from those of other seasonal variables such as nutritional and climatic stress. While there is evidence of an influence of day length, the other factors named are probably of greater importance in European breeds (*Bos taurus*) as well as in Zebu cattle (*Bos indicus*) and also in Water Buffalo (*Bubalus bubalis*). In Zebu cattle in the Mexican Gulf region the heifers are mated so that half of them conceive in the unfavourable dry spring season, but with successive calvings an increasing proportion conceive later in the year (Table 4.1).

Table 4.1   Percentage distribution of conceptions between wet and dry seasons in succeeding pregnancies of Zebu cattle in the Gulf of Mexico. (From Jochle, W. (1972). *International Journal of Biometeorology*, **16**, 131.)

	Parity			
	0	1–3	4–6	7–9
Dry season (Jan–May)	49	36	28	28
Rainy season (June–Oct)	51	64	72	72

Puberty

The age of puberty varies with breed and plane of nutrition (pp. 30, 39). Generally speaking, it is attained at about two-thirds of mature body weight and at less than one year of age in well fed animals.

Rate of sperm production by the bull increases gradually with age, and at above 12 months old is directly proportional to the size of the testes (Table 4.2). The first oestrus is considered to mark the moment of puberty in heifers, and this occurs earlier, and at lower body weight, in those born in spring (Roy *et al.*, 1977–78). However, considerable 'silent' ovarian activity (unaccompanied by oestrous behaviour) precedes the first heat; increasing pituitary activity causes development of follicles, and ultimately

Table 4.2   Development of sperm production in Holstein-Friesian bulls. (From Amann, R. P. and Almquist, J. O. (1976). *Proceedings of Technical Conference on Animal Reproduction and Artificial Insemination, Milwaukee*, **6**.)

Age	Number of bulls	Gross weight of testes (g)	Daily sperm production ($\times 10^6$)	
			Total	Per g testis
0–4 months	25	20	0	0
5–7 months	15	97	104	1
8–10 months	20	284	1750	7
11–12 months	15	370	3300	10
17 months	13	480	4480	10
3 yr	10	586	6040	11
4–5 yr	11	647	6530	11
> 7 yr	11	806	8000	11

one ruptures. The resultant corpus luteum commonly produces relatively little progesterone and has a brief period of function, so that a further ovulation occurs about ten days later. Thereafter normal ovarian cycles follow. Similar ovarian activity may precede the resumption of normal cycles in the cow after calving.

The oestrous cycle

Heats occur every 20 days in the heifer and 21 days in the cow, but these periods may vary 2 or 3 days either way. Much longer periods are often recorded, although how far these are due to short heat periods which are not seen is in doubt. The duration of heat averages about 18 hours (range 6 to 30 hours). Similar figures apply to the buffalo.

During each oestrous cycle there is a regular sequence of changes in the ovary (Fig. 4.2) which are reflected in changes in the reproductive tract. Before oestrus there is the development of a follicle associated with the production of oestrogen. This causes the cow to come into heat and stand for the bull. At the same time the oestrogen causes the cervix to dilate and its mucus becomes thin and slimy, so that it flows from the vulva. Bleeding may occur in the uterus, and sometimes blood-stained mucus appears at the vulva about 72 hours after heat. This occurs most commonly in heifers and in cows in good condition. If the time of the cow being on heat has been missed, its appearance can be used to predict the next heat. At the time that the cervical mucus becomes fluid, its electrical conductivity increases; its measurement is a possible way of determining when to inseminate animals failing to manifest oestrous behaviour—if suitable instruments can be developed.

Following ovulation, a corpus luteum develops in the ruptured follicle in a manner similar to that described for the mare. The duration of the cycle is controlled by the life of the corpus luteum. Ovulation occurs some 12 hours after the end of oestrus (Hansel and Trimberger, 1951). The fresh corpus luteum reaches its full size in about 8 days and persists in an active condition for a further 9 or 10 days—a total of 17 or 18 days. If the corpus luteum is removed (Fig. 4.3), as can be done by squeezing it out of the ovary by manual pressure through the wall of the rectum, the amount of progesterone in the blood draining the ovary falls very rapidly. The inhibition of release of the gonad-stimulating hormones is removed (see page 17), and a follicle will ripen and the cow will come on heat in 2 to 6 days. This is most easily done at the 8th to 12th day after oestrus.

Such treatment can be used as an aid to getting in calf animals which have not been detected in heat—both by advancing the time of the next ovulation, and by enabling one to predict when it will happen. Treatment with a suitable prostaglandin analogue (p. 15) will have the same effect— provided a corpus luteum is present in the ovary—and is preferable if the operator is not well practised. With either method, should the animal happen to be already pregnant, abortion will be induced (unless pregnancy is far advanced, when the placenta may produce sufficient progesterone to maintain pregnancy).

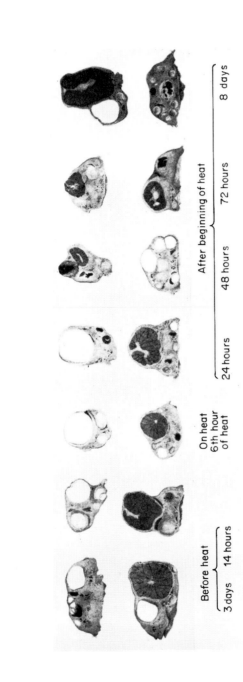

Fig. 4.2 Ovaries of cows at different stages of the oestrous cycle. The two ovaries (right and left) of an animal are shown, one above the other. The top row shows the ripening follicle which ruptures between 24 and 48 hours after the beginning (or 14 hours after the end) of heat and forms the new corpus luteum, which is at first dark from a blood clot (48 hours) and later becomes rather paler and increases greatly in size (8 days). In the bottom row, which shows the ovary with the old corpus luteum of the previous heat period, the stages of its degeneration are seen. (Hammond, J. (1927). *The Physiology of Reproduction in the Cow*. Cambridge University Press.)

Before heat

3 days   14 hours

On heat
6th hour
of heat

After beginning of heat

24 hours   48 hours   72 hours   8 days

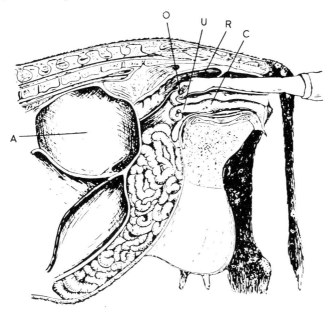

Fig. 4.3 Method of squeezing out the corpus luteum in the cow. The hand is inserted in the rectum (**R**). After the hard cervix (**C**) has been felt, the division of the uterine horns (**U**) can be felt on the brim of the pelvis and, following these round, the ovaries (**O**) can easily be detected and the protruding corpus luteum (see Fig. 4.2) squeezed out. (**A**) is the cavity of the rumen. (Marshall, F. H. A. and Hammond, J. (1952). *Fertility and Animal Breeding.* Ministry of Agriculture Bulletin, No. 39.)

Fertility and sterility

Twins are uncommon, though less so in dairy than in beef breeds, because the cow rarely (naturally) sheds more than one egg at each heat period. In British dairy breeds the incidence increases with the age of the cow, from about 1 in 150 in the first pregnancy to 1 in 30–50 by the fifth calving. Most twins are the result of the shedding of two eggs, but in about one pair in eight sets twins are identical, that is, both develop from a single fertilized egg. Identical twins are always of the one sex and are fertile. Non-identical twins may be male and female; in which case the female is usually infertile and is called a freemartin (see page 63); the male is fertile.

As with other species, the timing of mating relative to that of the shedding of the egg is important. However, because heat periods are usually short this is not the serious problem that it is in the mare. Best conception rates are obtained from inseminations done not less than 6 and not more than 24 hours before ovulation. This corresponds roughly to a period of 18 hours commencing about 6 hours after the onset of heat (Fig. 4.4) and extending beyond its end.

Both infective and physiological factors are involved in temporary sterility, and anatomical defects are involved in permanent sterility. Infective

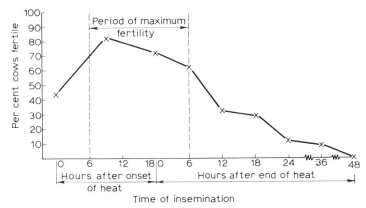

Fig. 4.4 Relationship between time of insemination and fertility of cows. (Drawn from data of Trimberger, G. W. and Davis, H. P. (1943). *Bulletin of the Nebraska Agricultural Experiment Station*, No. 153.)

conditions include contagious abortion, caused by the Bangs bacillus, and the venereal diseases Trichomoniasis (caused by a protozoan) and Vibriosis (caused by a bacterium). Contagious abortion is spread mainly by ingestion and is best avoided by immunizing, by inoculation of heifer calves with strain S19 at six months of age. Trichomonas and Vibrio foetus infections are spread by coitus from an infected bull. The use of disease-free bulls in the artificial insemination centres has virtually eliminated these conditions from large areas. Virus infections can also cause temporary sterility.

Sterility, mainly temporary, formerly accounted for about 25 per cent of the annual wastage of dairy cows in England. Elimination of venereal diseases, and of Brucellosis, has reduced this wastage, and of course tended correspondingly to increase the proportion disposed of for other reasons— the main ones being poor milk yield, mastitis and (happily) old age. But reproductive failure remains high among the list of reasons given for culling.

An intermediate condition—an interaction between infective and physiological factors—sometimes occurs. This is persistence of the corpus luteum, usually as a result of an infection of the uterus or the retention of a portion of the placenta from a previous pregnancy. The consequent stimulation of the uterus simulates a pregnancy and the corpus luteum does not regress. If it is caused to regress by prostaglandin treatment, progesterone is no longer produced; follicles enlarge and produce oestrogen which increases the muscular tone and contractions of the uterus, dilates the cervix, and causes the uterine secretions to be bactericidal instead of nutrient. These effects combine to clear up any infection and to rid the uterus of unwanted contents. Removal of the corpus luteum in normally cycling animals similarly helps to clear up uterine infections. If a severe infection (such as of Trichomonas) is present, the additional use of an injection of 25 mg stilboestrol will assist in clearing it up.

Many cases of temporary sterility, or 'repeat breeding', also occur when the cycles are quite regular and no infective condition can be found. Casida has summarized available information comparing normal heifers bred for the first time with repeat-breeder cows and heifers (Table 4.3). The two most important causes of trouble are failure of fertilization and early death of the embryo, commonly between the 16th and 25th day of pregnancy.

In normal matings, and in artificial insemination, the number of cows which conceive is rarely above about 70 per cent and is usually below this figure. Estimates of the percentage of ova fertilized range from 66 in cows of low fertility to 100 in cows of normal fertility. The discrepancies between these figures and the final calving percentages are due to early loss of fertilized eggs, or of the embryos, usually within the first month of pregnancy, or death of the foetus followed by abortion late in pregnancy. The most serious loss, in cows free from contagious abortion, is in the first month. This appears to be due to a defect in the uterus, caused probably

Table 4.3   Comparison of clinically normal repeat-breeder cows and heifers with first-service heifers. (Summary by Casida, L. E. (1961). *Journal of Dairy Science*, **44**, 2323.)

Per cent animals with	Normal first service heifer	Repeat breeder	
		Heifer	Cow
Abnormalities of reproductive organs	2.7	13.5	6.0
Failure of fertilization and early loss of eggs	21.7	35.3	39.3
Loss of embryos before 30 days	16.0	24.8	32.5
Normal embryos at 30 days	59.6	26.4	22.2

by an endocrine inadequacy. Fertility at service within the first two months of calving is lower than normal; this is possibly due to nutritional stress at the time of peak lactation, or may be related to the post-partum involution of the uterus.

Some sterility is associated with abnormal follicle development, usually accompanied by abnormal sexual behaviour; usually also there is swelling of the vulva and raising of the tail head. In the true nymphomaniac, which is almost always on heat, there are multiple follicles which fail to ovulate. This condition is similar to that in the rat treated with androgen (page 17). More common are animals with one or two large cysts in the ovaries. These cysts are follicles which fail to ovulate, but continue to secrete oestrogen. Their granulosa cells, and ovum, degenerate after a while, and their walls become tough. After rupturing such a cyst (per rectum) another cyst is often formed; but if this is broken within a week or so, or if it is induced to ovulate by injecting the animal with HCG, a corpus luteum forms from it, and normal heat and ovulation follow three weeks later.

A peculiar form of sterility occurs in heifers born twin to a bull. In about 90 per cent of such cases the foetal circulations fuse (Fig. 4.5) early in development, before the gonads have differentiated. The testis develops earlier than does the ovary, and a hormone which it produces suppresses — not only in the bull calf, but also in the heifer twin—development of ovarian elements and (to a variable extent) formation of the cervix and

Fig. 4.5 Twin Chorionic Vesicle of cow, showing anastomosis of the foetal circulation of differently sexed embryos. (After Lillie, F. R. (1932). In *Sex and Internal Secretions*. (E. Allen, Ed.) Bailliere, Tindall and Cox, London.)

uterus. In this respect the freemartin resembles 'white heifer disease' (p. 215); however the 'white heifer' has normally functioning ovaries. The freemartin appears effectively to be a castrate, but it seems that sometimes male germ cells can migrate into the ovary and cause partial development of testis tissue. At a later stage of foetal development, androgen passing from the bull calf to the heifer can cause some masculinization of the external genitalia; the clitoris may be enlarged and extend downwards (as the penis develops in the male).

A considerable amount of sterility may be due to lack of sperm production by the bull. This may be due to infections (see Fig. 1.5). It is said that poor sperm production is found in young bulls which have been over-used and under-fed, and in old bulls which have been over-fed and under-used. There is good evidence for the former statement but not for the latter. In the first ejaculate after a long period of rest there may be a large volume of accessory secretions and many dead sperm. With frequently repeated collections, the supply of spermatozoa falls more rapidly than that of

Table 4.4 Average calving interval in a herd of buffalo in Pakistan. (From Ashfaq, M. and Mason, I. L. (1954). *Empire Journal of Experimental Agriculture*, **22**, 161.)

Year	Calving interval (days)
1947	614
1948	528
1949	484
1950	436
1951	385

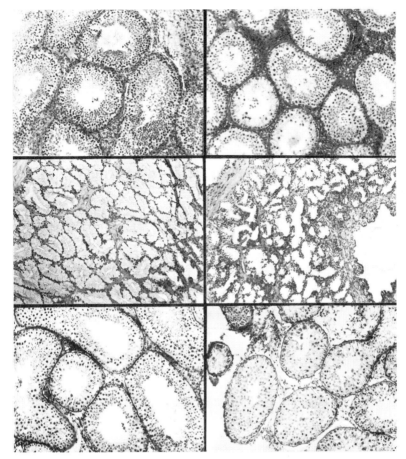

Fig. 4.6 Effects of nutrition on bull testis development. *Right,* low-plane feeding; *left,* high-plane feeding. *Above,* testis; *centre,* Cowpers gland; both at $10\frac{1}{2}$ months. *Below,* testis at $13\frac{1}{2}$ months. All to same magnification. In the identical twin calf on low-plane feeding, accessory secretory activity is much less well developed, and at $13\frac{1}{2}$ months (*lower right*) the testis tubules contain fewer cells at all stages of sperm formation. (Davies, D. V., Mann, T. R. R. and Rowson, L. E. A. (1957). *Proceedings of the Royal Society, B,* **147,** 332.)

accessory secretions, although very many more sperm remain in the epididymis. Such poor samples of semen are clear rather than milky in appearance.

The influence of under-nutrition on testis function is illustrated in Fig. 4.6. It was once thought that exercise was important, but experience in artificial insemination centres has not confirmed this. A normal well-balanced diet, coupled with sensible use of the bull, will result in good sperm production.

Buffalo are generally kept, either for milk or for work animals, where the climate is too extreme for cattle, or for horses and mules. In either case feeding conditions, environment and husbandry are usually sub-optimal, and reproductive performance (relative to that of cattle) reflects this. Thus Hafez (1955), in Egypt, reported average age at first oestrus as $13\frac{1}{2}$ months, but at conception $21\frac{1}{2}$ months. There is also a high incidence of silent heats and the interval between calvings tends to be prolonged; but this (see Table 4.4) can be much improved by improved nutrition and management.

Artificial control of breeding

There are four important techniques; control of the time of ovulation, control of the number of ovulations, transfer of embryos, and the long term storage of frozen embryos.

Controlling the time of ovulation

This may be necessary when it is required to inseminate animals at a particular time, or when it is necessary to synchronize times of oestrus and ovulation for the purpose of embryo transfer.

The alternatives exist either of shortening the cycle or of lengthening it. Corpus luteum function can be ended prematurely; or by treatment with progesterone, or a progesterone analogue, ripening of a fresh follicle can be delayed until some time after the corpus luteum has naturally regressed.

Expression of the corpus luteum from the ovary is only practicable in about the middle third of the cycle; subcutaneous injection of a synthetic prostaglandin (such as I.C.I. 80996, 'Estrumate') is simpler and quicker, and is equally effective, except in the earlier days of the cycle. If it is required to synchronize a group of animals all at different (and perhaps unknown) stages of the cycle, it is necessary to treat them twice, at an interval of about twelve days. Those that respond to the first treatment will then also respond to the second, and those refractory to the first injection will also be in a condition to respond on the second occasion. The timing of the subsequent ovulation is sufficiently predictable that insemination at a fixed interval after the second injection (without watching for the behavioural signs of oestrus) can give reasonable conception rates.

Table 4.5 Calving rates of Estrumate treated cattle and contemporary cattle bred between 18.11.75 and 31.12.75. (From *Milk Marketing Board Production Division Report, 1976/7,* **27,** 89.)

	Number of inseminations	Number calving	Calving rate
TREATED			
Dairy cows	1236	495	40.0%
Dairy heifers	3597	2081	57.9%
Suckler cows/heifers	770	309	40.1%
CONTEMPORARIES			
Dairy cows	4160	2138	51.4%
Dairy heifers	159	86	54.1%
Suckler cows/heifers	94	44	46.8%

Table 4.5 gives results of a trial by the Milk Marketing Board, in which animals treated in this way were inseminated either twice, at 72 and 96 h, or once at 72–80 h after the second injection. It will be noted that, in general, results were better with the control animals.

Progesterone, to delay ovulation, can conveniently be administered by placing into the vagina a plastic spiral device (Fig. 4.7) impregnated with the hormone. These are better retained than the plastic sponges (see Fig.

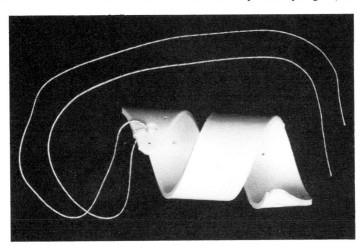

Fig. 4.7 Progesterone releasing intravaginal device (PRID) (Abbott Laboratories) for synchronizing oestrus in cattle.

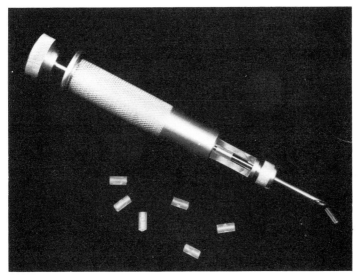

Fig. 4.8 Progestagen pellets and applicator (G. D. Searle) for synchronizing oestrus in cattle. Each pellet is enclosed in a cartridge of plastic tubing, from which it is expelled by the plunger of the syringe.

5.2) used for a similar purpose in sheep. Alternatively, a pellet of a potent progesterone analogue can be implanted under the skin of the ear with a special applicator (Fig. 4.8). Either method is conveniently combined with prostaglandin treatment, so that the progesterone need only be given for 10 or 12 days instead of for the 21-day duration of a natural cycle.

After removal of the pellet, or plastic spiral, about 90 per cent of treated animals can be expected on heat 72 to 96 h later. Some investigators have injected MSG, with the object of speeding follicle growth.

Controlling the number of ovulations

To obtain fertilized ova for embryo transfer it is an advantage to induce multiple ovulation; to get twin calves, a possible approach is to induce twin ovulations.

Injection of MSG causes growth of follicles which do not ovulate until the corpus luteum regresses, or is removed. If the dose is large, and the interval before the corpus regresses is fairly long, many eggs may be shed (Fig. 4.9); with lower doses, or shorter intervals, there are fewer ovulations.

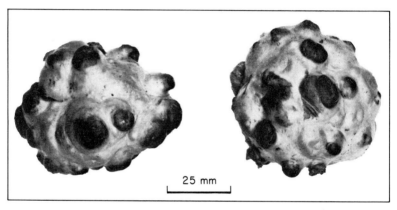

25 mm

Fig. 4.9 Ovaries of a cow which has been superovulated (51 eggs shed) by injection of 3000 i.u. of pregnant mare serum hormone 7 and 5 days before heat. (Dowling, D. F. (1949). *Journal of Agricultural Science*, **39**, 374.)

Excessive oestrogen production when there are numerous ripe follicles seems to be responsible for rapid transport of the ova to the uterus, where they cannot survive (page 11): but D. F. Dowling has obtained up to 25 eggs developing normally in the oviducts.

The major part of the cost of a beef calf is the keep of its dam for a year; production of twins would nearly halve the cost of the calves at birth. Figure 4.10 illustrates an early attempt at twinning. In attempts to obtain twin births, there have been great difficulties associated with unreliable control of the numbers of ovulations and with early loss of a high proportion of eggs. In a large-scale trial only 44 of 317 pregnant treated cows had multiple births.

In sheep—ovulating more than one follicle at a heat period—or animals

Fig. 4.10 Shorthorn dam and her three-month-old triplet calves born as a result of MSG injection before service. (Hammond, J. (1946). *Journal of the Ministry of Agriculture, London*, **53**, 34.)

such as pigs or ferrets, which shed many eggs, the ovulations seem to be distributed at random between the two ovaries. However the fertilized ova before implantation in these species move readily from one side of the uterus to the other, and generally become fairly equally distributed between the two horns. In the cow it would seem that the migration does not regularly occur, because few twin births were obtained from treated cows palpated as having two corpora lutea both in one ovary. Because of this, embryo transfer is now favoured as a method of obtaining twins.

A great many permutations of timing and dosage of MSG treatment—combined sometimes with prostaglandin and HCG and repeated inseminations—have been tried to find a method of consistently obtaining multiple fertilized ova for embryo transfer. So far the most satisfactory—providing, with luck, 8 or 9 embryos—represent a compromise between few ovulations, but good recovery of viable eggs, and many eggs shed, but unfertilized or of low viability.

The use of MSG has also been attempted to speed up breeding in cattle to obtain a generation a year. Thus Marden has shown that, with gonadotrophin treatment, it is possible to obtain fertilized eggs from calves a month old. The heifer calf starts life with some 75 000 eggs in her ovaries, while at the best a cow will not normally produce more than about 10 calves in a lifetime. Unfortunately, eggs obtained from pre-pubertal animals appear to be of poor quality, due either to excessively high levels of oestradiol and progesterone produced in gonadotrophin-treated heifer calves (Saumande, 1978), or to the lack of an essential maturation factor (Thibault, Gerard and Menezo, 1975).

Embryo transfer

Embryo transfer offers the possibility of twin production in beef cattle; it also provides a method for genetic improvement, on the female side, analogous to that provided by the development of artificial insemination. Much greater breeding use can be made of genetically superior females.

The pioneering work of Chang, with rabbits, demonstrated that fertilized eggs could be kept alive in blood serum and survive transplantation from one animal to another, provided that there was close synchrony between the times of ovulation in the donor and recipient. The reasons for this need for synchrony are partly the different nutritional needs of tubal and uterine stages of development, and the changes in uterine secretion during the cycle, and partly the limited period of time in which the uterus will respond to attack by the membranes to form a placenta.

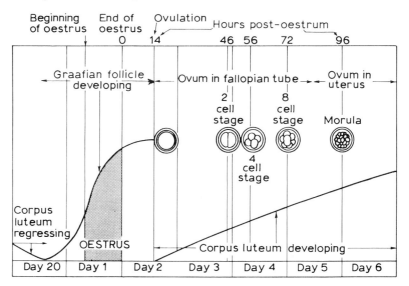

Fig. 4.11 Diagrammatic representation of events at and about oestrus in the cow. (Laing, J. A. (1949). *Journal of Comparative Pathology and Therapeutics*, **59**, 97.)

Figure 4.11 depicts the early life of the fertilized ovum. The egg loses its zona pellucida 2 to 3 days after it enters the uterus: as the membranes enlarge thereafter, the egg becomes more delicate and difficult to handle without injury. Figure 4.12 shows a device which can be passed through the cervix, and with which fertilized eggs have been flushed from the uterus 6 days after heat.

Transfers of eggs in cattle have been made by the same sort of techniques—and equally successfully—as those made in rabbits. But these involve major surgery, the expense of which is only justified if the value of the calves to be produced will pay for it (as indeed may be the case for some pedigree animals). There is the practical difficulty also—particularly if ova are recovered from the Fallopian tubes—that inevitable trauma may cause the animals to be subsequently sterile. For these reasons, methods are being developed for non-surgical recovery of ova from the uterus, and for non-surgical transfer to the uterus of the recipient.

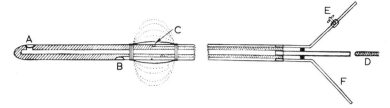

Fig. 4.12 Apparatus for washing out eggs from the uterus of the cow. The tube is inserted in the uterine horn. Air is admitted through tap **E** and opening **C** to inflate a rubber balloon which occludes the cavity of the uterus. The steel stilette **D** is removed and blood serum diffused through inlet **A** into the uterine cavity. This washes the eggs through outlet **B** and they are collected in a vessel held under tube **F**. (Rowson, L. E. A. and Dowling, D. F. (1949). *Veterinary Record*, **61**, 191.)

Early attempts to introduce eggs into the uterus through the cervix resulted in severe uterine infections—the uterine secretions that support the growth of the egg also supported the growth of accidentally introduced bacteria. Later attempts, in which stringent precautions avoided infection, failed; very possibly because the fluid which was introduced was expelled by the muscular tone of the uterus, and the ovum expelled with it.

In what seems to be the most successful non-surgical transfer method so far, a 5 to 9 day embryo, in a Cassou straw (see p. 75), is transferred to the host uterus (at the corresponding stage post-ovulation) by passing the

Fig. 4.13 A crossbred beef suckler cow with her second set of twins by egg transfer; the calf on the right is her own calf, the other developed from a 7-day old embryo non-surgically transferred to her a week after she was inseminated. (Photograph by courtesy of I. Gordon, University College, Dublin.)

insemination pipette through the cervix to near the top of the uterine horn (contralateral to the side on which the host's own egg has been shed, if twinning is the object of the transfer). Great care is required to avoid damaging the uterine lining during this procedure.

Figure 4.13 shows beef twins obtained in this way. To obtain suitable embryos for such transfers, it is only necessary to use a beef bull on suitable cattle going for slaughter, to kill them at a suitable interval after insemination, and to recover the embryos from the uterus at killing. But of course the process becomes more efficient if the donors are caused to shed several eggs, and the times of ovulation of donor and recipients have to be synchronized.

With genetically valuable animals as donors, a non-surgical method of recovery from the uterus, with repeated harvesting of crops of multiply-ovulated eggs is required. Quite good recovery has been reported with instruments which are refinements of that illustrated in Fig. 4.12. A flexible plastic catheter is introduced through the cervix and passed up to the tip of each uterine horn in turn. A cuff is inflated with air, to seal off the top part of the uterus, and fluid passed in through one channel in the catheter is drained off through another. The embryos settle at the bottom of the dish in which the fluid is collected.

Storage of embryos

As with sheep (see Fig. 5.5), cow ova will survive, and continue to develop when placed in the oviducts of rabbits; embryos can also be stored in a synthetic medium for 3 days at 5 °C.

The first successful freezing of embryos was achieved with mouse ova, and the first calf from a frozen embryo (Fig. 4.14) was born in 1973. The procedures are essentially similar to those for the freezing of semen (see p. 74). However the mass of the embryo, and the need to avoid destruction of many cells, pose special problems, and seven- to eight-day-old embryos are preferred for freezing.

Following collection, the embryo is placed in a medium containing either glycerol or dimethyl-sulphoxide. These substances are cryoprotectants—they replace some of the water inside the cells, and in some way protect the cells from damage during freezing. The medium is then cooled very slowly. At -5 °C crystallization is induced by adding a little frozen medium; this is to avoid supercooling. Slow cooling is continued to -50 °C, and then the temperature is brought more rapidly to that of liquid nitrogen (-196 °C) for storage. Thawing also has to be carried out slowly, and after thawing the cryoprotectant has to be displaced, because it is toxic at body temperature.

Freezing of embryos permits ready transport of genetic material from one country to another. Embryos frozen in New Zealand were flown to Australia in 1975, thawed there, and successfully transplanted to yield live calves (Bilton and Moore, 1977). This overcomes some problems of international health regulations, since embryos can be stored for longer than the incubation period of diseases (e.g. viruses) which the parents might have been exposed to, and the embryo perhaps be capable of transmitting.

Fig. 4.14 The first calf born following transfer of an embryo which had been frozen; with the cow in which it developed, at the Animal Research Station, Cambridge. (Photograph by courtesy of L. E. A. Rowson.)

Also, it offers the possibility of storing present-day cattle for comparison with those of future generations.

Artificial insemination

There are three ways of collecting semen from the bull for testing for fertility and for artificial insemination, namely by massaging the vas deferens through the rectal wall, by use of an artificial vagina, and by electro-stimulation.

The artificial vagina (Fig. 4.15) consists of a cylinder into which a rubber tube, like the inner tube of a motor tyre, is inserted and doubled over at

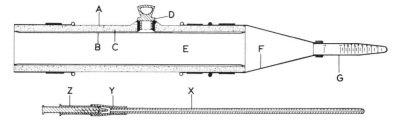

Fig. 4.15 (*Above*) Diagram of artificial vagina. A cylinder (**A**) fitted with a rubber tube (**B**) folded over the ends forming a space (**C**) in which water at body temperature is poured through the tap (**D**). The inner side of the rubber tube (**E**) is lubricated with white Vaseline. A soft rubber funnel (**F**) is bound over one end of the cylinder. Into the end of the funnel a graduated glass tube (**G**) is inserted. (*Below*) Apparatus used for insemination. A small (2 ml) glass syringe (**Z**) is fitted with a short piece of rubber tube (**Y**) into which is fitted a plastic tube (**X**) with a narrow lumen.

the ends. The space between is filled with air and water at deep body temperature. The artificial vagina is held alongside the flank of a cow or in a dummy so that the bull serves into it when he jumps (Fig. 4.16). The semen is ejaculated directly into the cone or collecting tube and not onto the rubber lining. Immediately service has occurred the artificial vagina is tipped up so that the semen falls to the bottom of the glass tube.

The electro-ejaculator is particularly useful for bulls which are old or have sore feet or some other disability which prevents their serving normally. Since the development of the initial type of apparatus there have been several modifications. Two types are commonly used today. One, developed by Rowson and Murdoch, consists of two copper rings used as electrodes attached to the fingers of a rubber glove worn to massage the vas deferens per rectum. The other is similar to the ram probe (see Fig. 5.6) and consists of an insulated rod with two circular electrodes. However, the procedure is liable to inflict pain, and on this ground the massage method is to be preferred.

The sperm are very sensitive to temperature and above 42 °C (107.6 °F) they soon die. Cooling slows down their activity and prolongs their life. Rapid cooling, however, damages them; slow cooling is essential. Certain substances protect the sperm against 'temperature shock' and are used in the dilution of semen. Dilutors include egg yolk with phosphate or citrate buffers, and skim milk or reconstituted dried milk powder. Many other substances and combinations have been used and antibiotics are commonly added.

The protective action of glycerol was used by Smith and Polge in 1950 for the freezing of bull sperm. Since these early experiments, 'deep freezing' of bull semen has become a standard practice. There are many variations in technique but the essential principles are dilution with a standard diluent and cooling to 5 °C, followed by addition of a cryoprotectant such as glycerol. The mixture is then allowed to stand for several hours to 'equilibrate'. Originally it was placed into sealed ampoules each containing enough for one insemination and cooled slowly to the temperature of

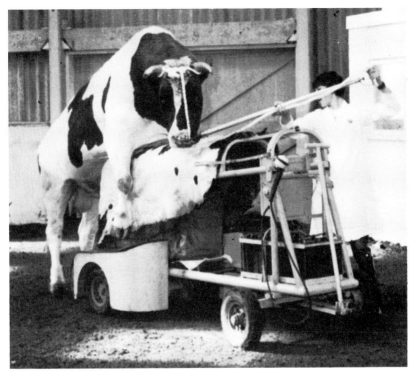

Fig. 4.16 Semen collection, using a mobile dummy cow. A technician con-
cealed within it holds the artificial vagina. (Photograph by courtesy of the Milk
Marketing Board.)

storage in liquid nitrogen ($-196\,^{\circ}$C). A more recent development is the
rapid freezing of the semen in a concentrated form in pellets (Nagase and
Niwa, 1964) or plastic straws (Cassou, 1968). Following equilibration, the
semen is either placed directly onto dry ice in small drops or is drawn up
into fine tubes (straws) which are lowered into liquid nitrogen (Fig. 4.17)
in which they are stored.

Work in New Zealand indicates that, to make maximum possible use of
an ejaculate, semen should be only partially diluted before freezing, dilution to the final concentration being done after thawing.

The standard method of insemination is to insert 1 ml of diluted semen
into the cervix, or through the cervix into the uterus, using a glass or
disposable plastic inseminating tube attached to a syringe (Fig. 4.15), or a
special pipette for semen stored in straws. The tube is guided by a hand
inserted in the rectum. Some centres prefer insemination into the cervix
rather than into the uterus, for fear of introducing infection or of the
remote possibility of damaging the membranes of an existing pregnancy.

By artificial insemination the number of cows that can be impregnated
from a bull can be greatly increased. Forty top bulls in Britain in 1975–76
produced an average of over 41 500 insemination doses of semen each.

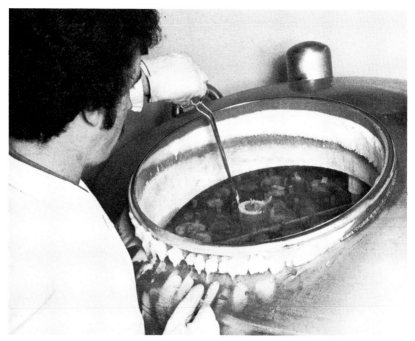

Fig. 4.17   A tray of straws being placed in a liquid nitrogen flask. (Photograph by courtesy of the Milk Marketing Board.)

Frozen semen can be kept in quarantine, for years if necessary, until the bulls from which it came have passed the incubation test for any disease. This has permitted the introduction into Australia and New Zealand of semen of cattle in the United Kingdom—notably the Charolais—despite restrictions in those countries on importations of livestock (Fig. 4.18).

Diagnosis of pregnancy

Absence of heat does not necessarily imply pregnancy, and pregnant animals sometimes show oestrous behaviour. The standard method of confirming pregnancy is by palpation of the uterus. Quite early in pregnancy membranes developing from the foetus extend round into the opposite horn of the uterus and, as fluid accumulates within the membranes, both horns of the uterus become enlarged. The foetus, however, is contained in fluid within an inner membrane, the amnion. The amnion at first is spherical, and when large enough it produces a localized distension in one uterine horn. In heifers this can be palpated as early as the 35th day— but care must be taken not to compress the swelling and so rupture the amnion. The cow usually has a larger uterus than the heifer, so that swelling can only be located with certainty a few days later than in the heifer.

Fig. 4.18 Junior Champion Guernsey cow at the 1975 Sydney Royal Show. The cow, Caronia Dari Pansy, was bred from frozen semen imported from England. (Courtesy of *Queensland Country Life*.)

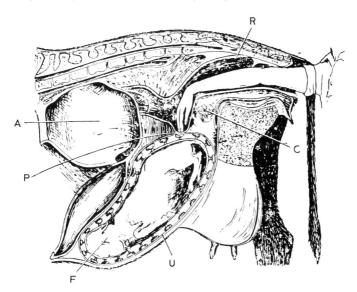

Fig. 4.19 Method of rectal examination for determining pregnancy in the cow. The hand is inserted in the rectum (R); the cervix (C) can be felt, but the uterus (U) and ovaries have dropped into the body cavity and only the lower end of the uterus and foetus (F) and a few of the cotyledons (P) can be touched. (A) is the cavity of the rumen. (Marshall, F. H. A. and Hammond, J. (1952). *Fertility and Animal Breeding*. Ministry of Agriculture Bulletin, No. 39.)

By about the 60th day the uterine swelling is no longer sharply localized. As the uterus continues to increase in volume it drops downwards into the body cavity. At this stage an infected uterus distended with pus might be mistaken for one that was pregnant, but as pregnancy advances the individual cotyledons (the multiple points of contact between the membranes and the uterus which make up the placenta in ruminants) can be felt (Fig. 4.19). The size of the cotyledons, and whether or not the calf itself can be felt, provide guides to the stage of pregnancy.

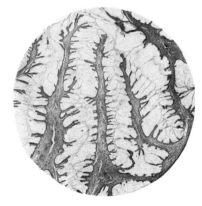

Fig. 4.20 Magnified sections through the cervix of the cow. (*Left*) Shortly after oestrus. The lumen between the folds is free from mucus. (*Right*) About $3\frac{1}{2}$ months pregnant. The lumen between the folds is distended with thick rubber-like mucus. (Hammond, J. (1927). *The Physiology of Reproduction in the Cow*. Cambridge University Press.)

During pregnancy mucin accumulates in the cervix (Fig. 4.20) and forms a seal which prevents the infection of the uterus from the vagina. As pregnancy advances the mucin becomes thick and the amount increases progressively: examination of the cervix and of the consistency of the mucus provides another guide. In heifers, and in cows which have been long dry, the secretion in the teats changes characteristically after the fifth month (see *Development of the udder*, below).

The corpus luteum persists in the ovary throughout pregnancy, and abortion generally follows its removal, except towards the end of pregnancy. The placenta of the cow produces progesterone, but relatively rather little. Oestrogen appears in the urine in increasing amounts, but does not rise significantly above non-pregnant levels until after the fifth month.

If a cow does not hold to service, the corpus luteum starts to regress at about 18 days, and the blood progesterone level falls. Following the next ovulation it rises again, but for 6–8 days it will be below the level normal in early pregnancy. Progesterone levels in milk reflect those in blood, and are generally rather higher. Concentration of progesterone in cows milk 24 days after insemination therefore indicates whether or not the animal is pregnant (Heap *et al.*, 1976).

The progesterone concentration in milk is measurable by a radioisotope method (p. 14) and this test has been automated and made available in Britain by the Milk Marketing Board. In 1977–78 over 100 000 animals were tested. Twenty per cent were found to be not in calf, 2.1 per cent gave doubtful positive readings (more often than not such animals prove to be in calf) and the remainder were diagnosed as pregnant. This very high figure of nearly 78 per cent pregnant is due partly to the fact that animals seen to come on heat before 24 days would not have been tested; in addition, a few would give false positives in the test because of having been originally inseminated at a time when they were not on heat. However about 15 per cent of those actually pregnant at day 24 apparently suffer later embryonic loss.

Parturition and birth weight

Gestation length is about 280 days for cattle and 315 days for buffalo, but varies with the breed, and with the genotype of the calf. The birth process is initiated by the calf, through its adrenal (p. 18) and placenta; but premature birth can be induced by treating the cow with large doses of glucocorticoid. Injection in late pregnancy of a long-lasting preparation of dexamethasone (an adrenocorticoid analogue) will induce calving, on average about 10 days later. This treatment has been widely used on dairy cows in New Zealand to shorten the calving season (Welch et al., 1977). If pregnancy is not sufficiently far advanced at the time of treatment, calf mortality is high, milk production is lowered and there is a high incidence of retained placentae.

Experimental crossing of large and small breeds (see Fig. 10.7), as in horses (see Fig. 3.7), showed that the size of the crossbred calf from the large cow was limited by its genetic growth potential—intermediate between that of the parents—while the calf from the small cow was kept small, presumably because of the small size of its placenta and limited nutrition received from its dam.

This observation encouraged the importation into Britain of the large Charolais beef cattle for crossbreeding with dairy cows (Chapter 12). Paradoxically, there were more calving difficulties with Friesians than with the smaller dairy breeds; in these the conformation of the pelvic canal is better adapted to the passage of the calf. However, when pure Friesian embryos were transferred to Jersey cows serious calving difficulties were encountered, particularly with bull calves. The probable explanation of the apparent discrepancy is that, though the small cow restricts the size of the calf it is carrying, the size of the head—being an early-developing part—is much less affected than total body size. Heifers are normally mated well before body size—and consequently capacity of the pelvic canal—is fully developed; it is therefore inadvisable to crossbreed them with a larger breed.

Development of the udder

The udder forms in the foetus by a down growth from the skin, interacting with the underlying connective tissue. This rudiment then develops further, under the influence of ovarian and pituitary hormones. In many species—almost certainly including cattle—androgen from the foetal testis impairs the capacity of the rudiment to respond to such stimuli.

In the calf the downgrowth hollows out to form the milk cistern, and from this ducts branch out to form the beginnings of the gland. At birth these are limited to a small area just above each teat, and can be felt through the skin. In the first few months after birth the amount of gland tissue grows rapidly (Fig. 4.21). Swett believed that the extent of gland

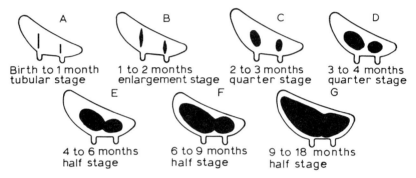

Fig. 4.21 Mammary gland development in Holstein and Jersey heifer calves and standards for evaluation. Data obtained in herd of Bureau of Dairy Industry Research Administration, U.S.D.A. Stages of development and approximate ages at which they normally occur. (Swett, W. W. *Year Book United States Department of Agriculture 1943–1947*, 195.)

development when the calf is four months old reflects the capacity of the udder to grow and secrete under the hormonal influences under which it will further develop when adult. This has not been confirmed.

After the age of about four months the deposition of fat in the udder makes accurate assessment of the amount of gland tissue difficult. This fatty tissue is eventually displaced by the development of gland in the latter part of pregnancy. During repeated oestrous cycles the duct system gradually branches and extends through the fat pad, under the influence, primarily, of oestrogen (Fig. 4.22). A little opalescent watery secretion can be drawn from the teats in the virgin heifer.

During the first five months of pregnancy there is little further development. At the 20th week of pregnancy, however, the cells of the alveoli begin to grow, and secrete a sticky fluid rich in globulin. As mentioned above, presence of this honey-like secretion in the teats can be used as an indicator of pregnancy.

This is a critical stage in udder development: abortion at (or after) this stage will be followed by a lactation—but the yield will be small because

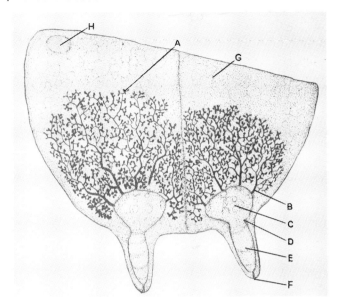

Fig. 4.22 Diagram of a cross-section of the udder of a 2 year old maiden heifer. **(A)** Alveoli which will eventually produce milk when they have been developed by pregnancy; these open into small ducts which join up to form large ducts **(B)** leading into the milk cistern **(C)** which acts as a reservoir for the milk as it is formed. Between the cistern and the cavity of the teat **(E)** there is a constriction **(D)** which is liable to be closed when the cup of the milking machine climbs up towards the end of milking. A narrow canal **(F)** with sphincter muscles closes the end of the teat. In the heifer a large part of the udder consists of fat **(G)** in which at the hind end is embedded a lymphatic gland **(H)** which filters the lymph coming from the udder and so swells up when the udder becomes infected.

the udder development is far from complete. As pregnancy advances there is further growth of the secretory part of the gland (the lobule alveolar system) with replacement of fat and accumulation of more secretion (Fig. 4.23). Dilution of the honey-like secretion with milk provides the colostrum. The colostrum is rich in antibodies, which the calf absorbs unchanged during the first two-days of postnatal life—and so acquires passive immunity from its dam. Primarily, the earlier stages of development are due to ovarian steroids and later stages of development to the pituitary hormone prolactin and presumably the 'lactogenic' hormone secreted by the placenta. The increasing levels of circulating steroid at the fifth month probably enables the gland to respond to prolactin. At this same stage of pregnancy the yield of a concurrent lactation becomes progressively depressed (see Fig. 4.29); it seems that at this stage the gland begins to be switched from secretion to growth in preparation for the lactation to follow.

It must be remembered however that the availability of nutrients to the different tissues is affected by a variety of metabolic hormones (page 25) and that the mammary gland develops in competition with the growing

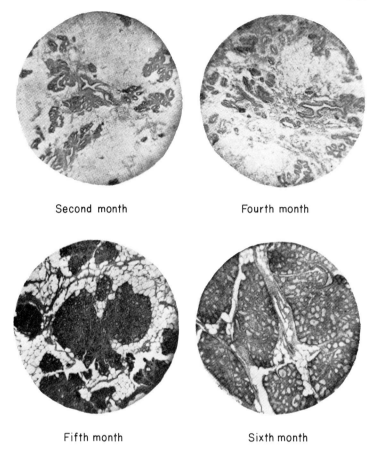

Second month   Fourth month

Fifth month   Sixth month

Fig. 4.23 Magnified sections through the udders of heifers pregnant for the first time. Up to the fifth month the udder consists largely of fat, but at this time the alveoli begin to grow and replace it. By the sixth month a considerable amount of secretion has appeared in the alveoli. (Hammond, J. (1927). *Physiology of Reproduction in the Cow*. Cambridge University Press.)

foetus, and, in the case of the heifer that is not full grown, with the other tissues. In the pregnant lactating cow there must be competition within the gland between formation of secretion, and development of the gland in preparation for the lactation to follow.

The amount of the alveolar cell growth which occurs during the latter part of pregnancy determines the amount of milk that can be given after the cow calves. This is the reason for 'steaming up' a cow before she calves (see page 91): poor feeding in the later stages of pregnancy leads to poor milk yields, especially in heifers. Milking up to the time of calving will reduce the yield in the subsequent lactation. As the dry period increases, up to about 50 days, the yield of the next lactation increases.

Udder development can be stimulated artificially, in the absence of

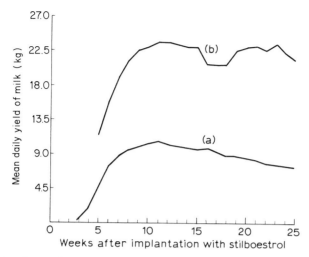

Fig. 4.24 Yields of milk resulting from treatment of **(a)** heifers with stilboestrol (Day, F. T. and Hammond, J. Jr. (1945). *Journal of Agricultural Science*, **35**, 151); and **(b)** an outstanding Friesian cow with progesterone + stilboestrol, followed by stilboestrol (Meites, J. (1959). In *Reproduction in Domestic Animals*. (H. H. Cole and P. T. Cupps, Eds). Academic Press, New York and London).

pregnancy. Maiden heifers implanted with tablets of synthetic oestrogen rapidly develop udder tissue and generally come into milk within about three weeks (Fig. 4.24); occasionally in the treated animals the corpus luteum persists, and while it does so milk yield is very low. Cows in milk treated in this way rapidly dry off—their milk becoming thick and sticky like colostrum—and rarely come into milk again during, or at the end of, treatment.

There are obvious parallels between these observations and what is seen in natural udder development. It is interesting that the corpus luteum, whether of the cycle or of pregnancy, seems to have no influence on established lactation, yet it appears to suppress the initiation of lactation, which generally begins as the corpus luteum dies away (and the placental secretions are withdrawn) at the end of pregnancy.

In the oestrogen implanted heifer, following removal of the implant after 60 days, the yield continues to increase for a while and then falls off, just as in a normal lactation. With this treatment yields up to 14 kg a day are obtained but there are undesirable side-effects. The sacroiliac ligaments relax, so that the tail head is elevated, and there is nymphomaniac behaviour; unless the animals are separately penned this combination of effects can lead to broken pelvic bones.

Greater yields (Fig. 4.24) have been obtained by treatment with a combination of oestrogen and progesterone (which suppressed side effects) followed by oestrogen alone. However, yields still remain uncertain, but clearly there is promise of a cheap method of treatment by which, at short notice, lactation could be started in a heifer.

More recently, Smith and Schaubacher (1973) treated nine barren cows, which had come to the end of their lactations, for a period of only seven days, with a mixture of oestrogen and progesterone, and seven of them came into milk again. Yields were 60-100 per cent as good as previous best lactations. In such a short time, it hardly seems possible than many new secretory cells were formed; rather, perhaps, the treatment rejuvenated those cells already present.

Lactation

In established lactation, injections of anterior pituitary extracts can increase the rate of milk secretion (Fig. 4.25); but this action is not brought about by prolactin—which may be necessary for maintenance of the gland, but clearly does not regulate its rate of secretion. Purified growth hormone will increase yield, but the effect may well be on the supply of precursors rather than by a direct action on the gland cells.

The administration of preparations with thyroid hormone activity can increase milk yield and rate of secretion of butterfat. The thyroid hormone is relatively inexpensive, but its effect on milk yield is not likely to be commercially useful because the effect of the hormone is to increase the metabolic rate of tissues generally—and over-dosage can be dangerous. Thus increased metabolic activity of the udder is matched by equally increased activity of the rest of the body, so that increased yield is not matched by increased efficiency of conversion of food into milk.

Milk secretion and composition

The mammary gland evolved from the apocrine sweat gland. Apart from

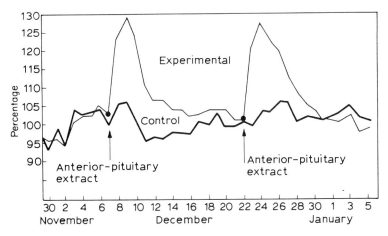

Fig. 4.25 Effect of injections of anterior-pituitary extract on the milk yields of cows, expressed as a percentage of the average milk yield during the preliminary period; the upper curve is for the experimental cows which received the extract, the lower curve is for the control group which did not receive the extract. (From Asimov, G. J. and Krouse, N. K. (1937). *Journal of Dairy Science,* **20,** 289.)

water and salts drawn directly from the blood, the main constituents of milk are synthesized within the secretory cells, which are those lining the alveoli and small ducts. Milk fat in general is closely related in composition to that of the ordinary adipose tissue, but in ruminants it also has a large fraction with shorter chain fatty acids, synthesized by the gland from the two and four carbon acids absorbed from the rumen. The carbohydrate is lactose and the protein mainly lactalbumins and the phospho-protein casein. Figure 4.26 diagrammatically illustrates the secretion process. Milk

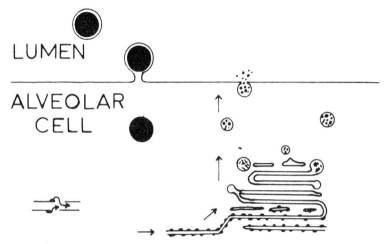

Fig. 4.26 Schematic representation of the two principal mechanisms of milk secretion. (*Right*) Synthesis and 'packaging of skim milk components into vesicles which fuse with and empty through the plasma membrane wall of the cell (merocrine secretion). (*Left*) Fat globules synthesized within the cell migrate to the wall where they are enveloped by the plasma membrane which is then pinched off (apocrine secretion). (Patton, S. (1978). *Journal of Dairy Science*, **61**, 643.)

has the same osmotic pressure as blood and osmotic pressure is largely determined by the concentration of small molecules; so there is a reciprocal relationship between the concentration of lactose and of chloride and other ions.

Figure 4.27 presents (for the goat) a balance sheet for materials removed from the blood and appearing in the milk. The processes of synthesis and secretion require energy, which is derived by the oxidation of glucose and of acetate. Lactose (and glycerol) can be accounted for by the other glucose removed. The mammary gland may account for as much as 70 per cent of the total body consumption of glucose. The protein secreted is accounted for by the amount of amino acids which are removed from the blood passing through the gland, and the rate of blood flow closely parallels the rate of milk secretion—so that the efficiency of extraction of milk precursors appears not to change with the rate of secretion (Bickerstaffe, Annison and Linzell, 1974). These authors provide a rather more detailed statement for the cow of the kind of information given in Fig. 4.27.

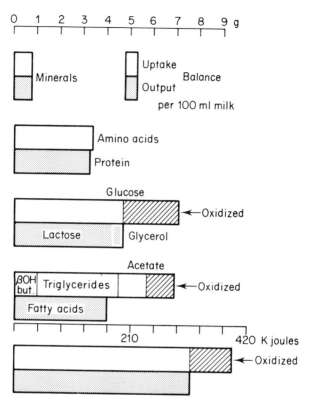

Fig. 4.27 Balance sheet of metabolic activity of the goat's udder during the secretion of 100 ml milk. The output of the major constituents is compared with the uptake of their chief precursors from the blood. The diagram ignores the uptake and output of minor components and the small degree of overlap between major components. The number of joules oxidized is calculated from the O_2 uptake, and the proportion of glucose and acetate oxidized is calculated from $^{14}CO_2$ production in isotope dilution experiments. β OH but. $= \beta$-hydroxy-butyrate. (Linzell, J. L. (1968). *Proceedings of the Nutrition Society*, **24**, 44.)

Many factors affect the composition of milk. Most important are the breed of cow and its feeding, and the stage of lactation. Generally speaking, a high fat content is correlated with high solid not fat (SNF)—protein (mainly casein), milk sugar (lactose) and minerals. Also there is an inverse relationship between total yield and per cent of fat (see Fig. 12.15) and of total solids: the milk of high-yielding Friesian cattle generally has a lower percentage of fat and SNF than that of the lower yielding Jersey. Buffalo milk has a much higher fat content and of solids not fat than that of cattle. At the peak of lactation the percentage solids is lower than at the beginning and end. The first drawn milk has a lower percentage of fat (about 2 per cent) than the last drawn (about 10 per cent) as the fat globules are much more held up in the fine ducts than is the SNF. During milking it is not possible to draw off all the milk from the fine ducts and usually some 6 to

20 per cent remains in the udder. This milk is high in fat, so that if the milk intervals are very uneven the fat percentage is higher after the shorter interval—usually the evening milking.

The butterfat secreted is derived partly from body fat withdrawn from the blood, and partly from fat manufactured within the udder from volatile fatty acids (acetic, butyric, etc.) from the blood. Acetic acid also serves, like glucose, as a source of energy. The tendency for fat percentage to fall with advance of lactation (but while yield remains high) is probably due to depletion of body fat reserves. The volatile fatty acids in the blood derive from ruminal digestion of fibre. A sudden change of diet, which alters the fermentation in the rumen, can cause marked transient drop in fat content—for example when high levels of starchy concentrates are fed, or when cattle are put out to fresh spring grass. The nature of the fat—and hence the 'spreadability' of the butter made from it—can be significantly modified by the inclusion of small amounts of different types of fat in the diet.

Milk formation proceeds, initially, at a constant rate during the interval between milkings: but after a while, as milk accumulates in the udder, a back-pressure will develop which slows the rate of further milk secretion. The interval after which this effect appears depends on the individual cow, and upon the stage of lactation. With a high-yielding cow, greater yields will be obtained by milking three, or even four times daily. However, with increasing labour costs, and the tendency to larger dairy units, the extra yield obtained is not generally considered worth while.

It sometimes happens that cows have a little extra, or accessory, gland in their udders. This is not milked out, and at first becomes very hard with the pressure of the accumulated milk. But within a few days it becomes soft again and no more milk is formed in the acessory gland. Continued secretion depends on removal of the milk formed, and it may be that back-pressure causes partial regression in a quarter that is milked.

Milk ejection or 'let down'

As milk is formed between milkings, some of it passes into the large ducts, and thence into the cistern and cavity of the teat (see Fig. 4.22): this milk is easily withdrawn at milking. The rest of the milk accumulates within the alveoli and fine ducts within which it is formed—the alveoli distend and the cells lining them become thin and flattened.

For this milk to be obtained, the cow has to 'let it down'. The phrase implies a passive process, but actually there is active expression of the milk from the alveoli into the larger ducts. This ejection of milk is brought about by contraction of muscular (myoepithelial) cells which surround the alveoli and small ducts (Fig. 4.28). The myoepithelial cells are not supplied with nerves: they are caused to contract by reflex discharge of oxytocin from the posterior pituitary.

The reflex is normally elicited by sensation from the teats, either by sucking of the calf or wiping with an udder-cloth. The reflex may also become a 'conditioned' reflex; that is, the cow may let-down her milk in

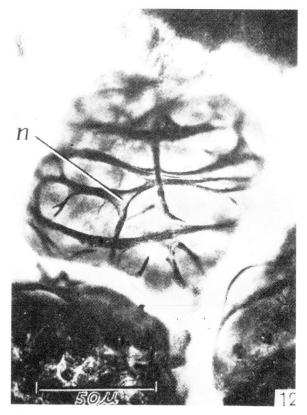

Fig. 4.28 A small contracted alveolus in surface view showing a myo-epithelial cell with nucleus (n) and branching processes. ($50\mu = 50\,\mu$m). (Richardson, K. C. (1949). *Proceedings of the Royal Society, B,* **136**, 30.)

response to some sound, or other stimulus, that has become associated with the act of milking. In the early stages of domestication (see Fig. 12.17) it was the practice to elicit the reflex by allowing the calf to suck before hand milking was begun. This figure shows reflex oxytocin discharge being stimulated in another way—by distention of the vagina (see page 6); one of the milkers is blowing into the cow's vagina. The reflex can be blocked centrally (in the brain) by alarming the animal, or by injecting the hormone adrenalin; stimulating the teats then causes no let-down.

The oxytocin released does not long remain in the blood, and a further reflex discharge cannot be obtained for a considerable time; milking-out must therefore quickly follow let-down. Table 4.6 shows how even a delay of four minutes can reduce the amount of milk that can be withdrawn. What is not withdrawn must remain in the udder, and will (by increasing the back-pressure which develops) perhaps reduce the milk available for withdrawal at the next milking.

As long ago as 1889, Babcock showed that slow hand-milking reduced milk and butter fat yield. Fast milking by machines does much to reduce

Table 4.6 Effect of delaying the interval from washing the udder to application of the teat cups on the amount of residual milk. (From Elliott, G. M. (1961). *Journal of Dairy Research*, **28**, 209.)

Treatment of half udder	Length of delay of half udder in minutes					
	0	2	4	6	8	10
			(kg of residual milk)			
Delayed	0.5	0.5	1.0	1.1	1.1	1.4
Control—not delayed	0.5	0.5	0.8	0.7	0.8	0.6

the liability of udder tissues to injury, and cows which inherently are fast milkers tend to be better producers than slow milkers. Hence today there is a strong trend towards standardized milking times, as short as four minutes, with no hand stripping. However, fast-milking cows are more susceptible to mastitis than are slow-milking cows due to the more open teat canals.

Management and milk yield

Under British conditions it is considered that optimal returns are obtained if cows calve at one year intervals. As after about 20 weeks of pregnancy yields are depressed (Fig. 4.29), the first 305 days is used as the standard period for lactation yield; so largely avoiding the need to consider the service period (interval between calving and following conception) in making comparisons. However, many other factors also affect yield.

In a normal lactation the yield rises for about six weeks after calving and then declines; in heifers the peak is less marked than in old cows, but their yield also drops off less toward the end of the lactation.

The shape of the lactation curve varies with feeding and other environmental conditions. In cows for which feeding conditions are poor the lactation curve does not rise during the first six weeks after calving but falls from the very outset. The shape also varies with the month of calving: the autumn-calving cows fall steadily in yield throughout the winter but in spring rise again due to being turned out on young grass rich in protein and minerals and low in fibre. Spring-calving cows, on the other hand, are dry before the next young grass season occurs and so do not show this lift in yield in the middle of their lactation (Fig. 4.30). Thus autumn- and winter-calving cows usually yield about 450 kg more milk in the lactation than do spring and summer calvers. This is important also in baby beef production. Calves born in autumn are better able to use the available milk supply and to use the 'spring flush' of feed and so grow better, and can be marketed earlier, than those born in the spring.

Environmental conditions at different seasons of the year account for large differences in milk output. Figure 4.31 shows the relative output of milk in a herd where equal numbers of cows are calving in each month of the year; if an even all-the-year-round supply is required, it is necessary to calve a higher proportion of cows during the winter months. Young grass accounts for the steep rise in yields in the spring. However, this rise

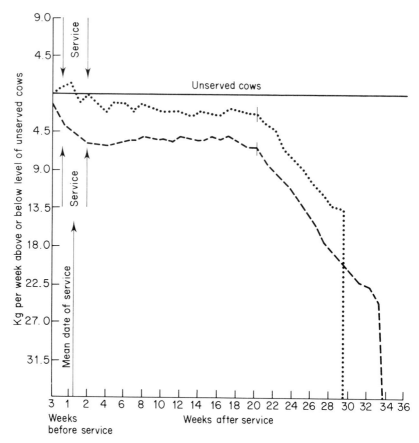

Fig. 4.29 Effect of pregnancy on the yield of milk in cows already milking. Note the sharp drop in yield, compared with non-pregnant cows, beginning at the 20th week after service. It is at this time that the mammary cells begin to grow in preparation for the next lactation. – – – – Norfolk; · · · · Penrith. (Sanders, H. G. (1927). *Journal of Agricultural Science*, **17**, 502.)

commences in mid-winter, and indeed precedes the increase in hours of daylight. It is possibly due to a photoperiodic effect on anterior pituitary hormone secretion.

Figure 4.32 illustrates an experiment on the effect of nutrition on lactation yield. Over the first 18 weeks after calving, some animals were fed on a constant high plane, some on a constant low plane; others, after 9 weeks, were changed between the two levels of feeding. When the level of feeding was reduced, milk yield fell—but not to the level in those which had been poorly fed from the start; and when those previously poorly fed were put on a higher plane their milk yield rose. But not all the extra food was returned as milk, a lot went to increasing body weight; in total yield over the 18-week period, those on the High–Low sequence gave more milk than

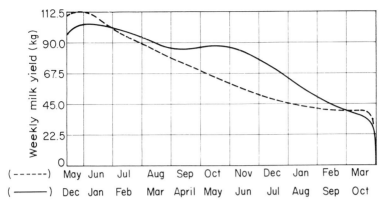

Fig. 4.30 Differences in the shape of lactation curves of May and December calvers. Although the May calver starts at a higher level, the December calver gives about 450 kg more because of the rise which occurs when it is turned out to grass in April (Northern Hemisphere).

those in the Low–High group. So perhaps the amount of gland secreting was diminished in the period of poor feeding early in lactation.

The yield is considerably affected by the plane of nutrition in late pregnancy, before lactation begins; it is therefore good practice to 'steam up' a cow or heifer by feeding a high-energy concentrate diet in the 6 weeks before calving. In such 'steamed up' cows it is sometimes necessary to begin milking before calving because of the build up in udder pressure (Hammond, 1936). As the volume of the uterus and its contents expands

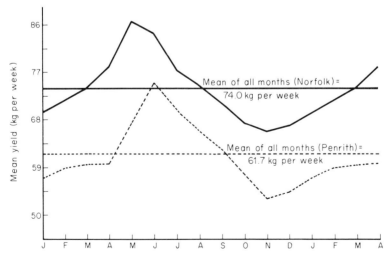

Fig. 4.31 Variations in milk yields during different months of the year in two different districts. Calculation based on the assumption that an equal number of cows calve during each month. (Sanders, H. G. (1927). *Journal of Agricultural Science*, **17**, 337.)

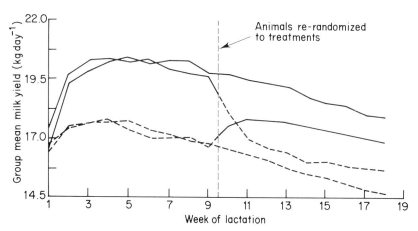

Fig. 4.32 Mean daily yields (kg) of milk of groups of heifers receiving one of two fixed levels of food in weeks 1–9 of lactation (treatments He, Le) followed by one of the same two levels in weeks 10–18 (treatments Hm, Lm). (——, He, Hm; – – –, Le, Lm) (level H > level L). (Broster, W. H., Broster, V. J. and Smith, T. (1969). *Journal of Agricultural Science*, **72**, 229.)

in pregnancy there is some compensatory reduction of gut capacity. This occurs mainly in the rumen; so the cow's capacity to ferment roughage diminishes, and its appetite falls toward the end of pregnancy—and only slowly recovers after calving. In the first few weeks of lactation the cow is drawing upon the reserves laid down in pregnancy; its output as milk exceeds its intake as food.

At each successive pregnancy after the first, a little more gland tissue is formed and so the milk yield rises up to about the fifth calf, although the

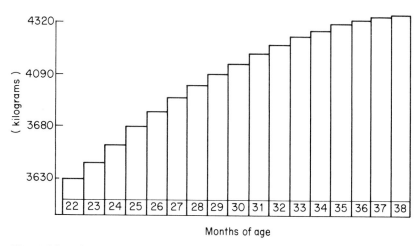

Fig. 4.33 Effect of age at calving on first lactation yield, corrected for season of calving and management; based on records of 160 000 Friesians. (Milk Marketing Board (1972–73). *Breeding and Production Report*, **23**, 79.)

increases after the third calf are small. On the average, there is an increase of about 30 per cent in milk yield from the first to the third calf, but this can be reduced considerably if the heifer is 'steamed up' well before her first calf.

In the heifer, the udder must be considered as being in competition with other growing tissues for the available nutrients. Lactation yield at first calving increases with age (Fig. 4.33); but this is not an argument for delaying service; the younger the age at calving, and the greater the number of lactations, the larger the return the animal gives (see Table 12.10). However, it is necessary that the young animal be well fed, as impairment is otherwise to be expected; late calving may possibly result in larger ultimate body size.

High-plane feeding in the earlier stages of life does not promote high milk yield. Customarily, dairy calves are not too well fed, and the experiment of Table 4.7 strikingly illustrates the wisdom of this practice. Control

Table 4.7   Effect of plane of rearing on subsequent milk yield. (From Little, W. and Kay, R. M. (1979). *Animal Production*, **29**, 131.)

Plane of rearing	Age (weeks) at mating	Weight (kg) at mating	Corrected lactation yields			
			1st	2nd	3rd	4th
'Barley-Beef'	43	302	1959	2918	3545	3210
'Barley-Beef'	78	443	2450	3216	3310	
Conventional	78	353	3863	4694	4813	

heifers were reared normally, others in the manner of 'barley-beef' animals; some of the latter were mated at the same age as the controls, some earlier. All were fed in the same way after first calving. Yield increases with age and parity in all groups—but the third (and fourth) lactation of the 'barley-beef' animals is less than that of the first lactation in the controls; as in McCance's rats (p. 31) early nutrition has a lifetime effect.

Two other factors apparently influencing yield deserve brief mention. Cows giving birth to twin calves have slightly higher yields than contemporaries (Wood, 1975); the effect is possibly to be accounted for by the twin placentae producing more placental lactogenic hormone, and thus stimulating greater mammary growth. For the second factor it is less easy to find a plausible explanation; it seems (Taylor *et al.*, 1978) that genotype of the calf affects the subsequent yield of its dam, and that the better the bull is, genetically, for milk the worse is the effect of the calf it sires.

Climate

Efficiency of food conversion, whether for growth or for milk production, depends upon the animal being adapted to its environment. Under too-cold conditions, maintenance requirements are increased; when it is too hot, food needed for maintenance is not decreased, but the animal decreases its heat production by reducing food intake (page 23): a high-yielding animal necessarily has a high metabolic rate.

Wild species, under natural selection, are adapted to their environment: so too are the various breeds of cattle to the climates of the districts in which they originated. There are various generalizations relating type of animal to geographical origin (Wright, 1954). Animals originating further from the equator tend to be more compact in build and larger in body size; but mountain breeds tend to be small. In hot humid regions cattle are small and darkly pigmented; in hot dry ones they are larger and pale in colour (Fig. 4.34).

The thermal capacity of dry air is low: clothing, or a thick coat of hair, provides insulation against heat loss by keeping a layer of still air next to the skin. In heat loss to the atmosphere, air movement—due to wind or to convection currents—is thus much more important than direct conductive loss. The advantage of a long coat, such as that of Highland cattle, is supposed to be that it channels off rainwater, so that it drips from the coat and does not penetrate and wet the insulating layer of shorter hairs beneath. The energy required to warm water is very much greater than that needed for a corresponding volume of air; and the energy required to get rid of the water by turning it into water vapour is very much greater again.

When considering the replacement of an indigenous breed by importing an improved one, the likely suitability of the chosen breed can be assessed by preparing climographs (Fig. 4.35) in which, by plotting mean monthly air temperatures and humidities against each other, some aspects of climate can be diagrammatically described. Similar diagrams may be made, based upon rainfall. In seeking to limit the effects of cold environment, the obvious steps are provision of shelter from rain and of wind breaks to moderate wind-speed. High-plane feeding lowers the critical temperature (see Fig. 2.1) by increasing heat production: it will also result in deposition of more subcutaneous fat, and so provide better insulation. Webster (1976) considers that British breeds, under British conditions, are rarely obliged to increase metabolic rate to maintain body temperature. The exceptions, requiring some environmental protection, are young calves (with large surface area, poor skin insulation, and no extra heat from ruminal fermentation) and older animals on only maintenance feeding exposed to driving rain (that is, with evaporation at the skin surface enforced by high air movement). Milk yield, however, suffers under less severe cooling than this, in spite of greater ruminal heat production. As the animal reduces heat loss by reducing the rate of blood flow through the skin, and as the mammary gland is derived from skin, it may be that cooling reduces milk secretion by reducing blood flow through the udder.

Several factors are involved in adaptation to high temperatures. Small size results in relatively large surface area (and hence increased heat loss by conduction): this is probably one reason for the relatively good heat tolerance of the Jersey, among British breeds. Zebu cattle (Fig. 4.34) are distinguished by having large skin folds, and large ears—which increase their surface area. They also have subcutaneous fat localized in a shoulder hump: in the absence of an insulating fat layer elsewhere, body heat should be able to reach the skin surface by conduction, instead of requiring to be pumped there by increased skin blood flow.

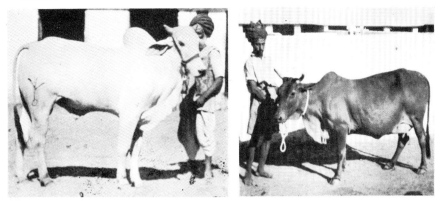

Fig. 4.34 Differences in the size of Zebu cattle which have been developed under different environmental conditions. (*Left*) Nagure bull from the dry area of Rajasthan. (*Right*) Hill cow of Red Sindhi type from the mountain area of Las Bela in Baluchistan. (Olver, A. (1938). *Miscellaneous Bulletin. Indian Council of Agricultural Research*. No. 17.)

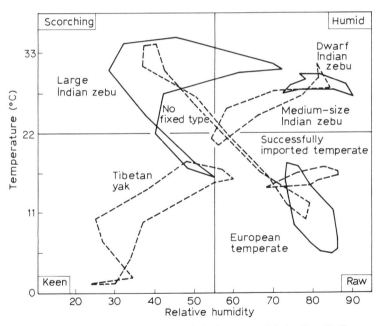

Fig. 4.35 Typical climographs for cattle in Europe and Asia. *Details:* European temperate (London, England); large Indian zebu (Delhi, India); dwarf Indian zebu (Cochin, southern India); yak (Lhasa, Tibet); successfully imported temperate (Nuwera Eliya, Ceylon); medium-sized Indian zebu (Karachi, Pakistan); no fixed type (Baghdad, Iraq). (Wright N. C. (1954). In *Progress in the Physiology of Farm Animals*. (J. Hammond, Ed.) Butterworth, London.)

Increased heat loss by sweating, and by increased frequency (with decreased depth) of respiration, become of greater importance at higher environmental temperatures. Many workers consider the greater heat tolerance of Zebu cattle to be due to a greater capacity for sweating. There is no doubt that the short sleek coat of the Zebu is an advantage, because clipping of the coat has been shown to increase the heat tolerance of British cattle in a tropical environment. Tropical cattle also are generally less productive—and hence have a lower feed intake and lower heat production.

Showering an animal provides it with a sort of artificial sweat to evaporate; the cooling effect can lower the body temperature for some hours. This effectively is what the buffalo does when it wallows; and clearly it augments the cooling effect to be obtained by sweating alone. Mehta *et al.* (1979) found that lactating buffalo that were either showered, or allowed to wallow, for half an hour twice during the day had shorter service periods and better conception rates than control animals.

Apart from high air temperature, solar radiation can impose an extra burden. Most of the ultra-violet is absorbed in the atmosphere (but can be harmful—see cancer-eye, page 239). Most of the radiant energy is almost equally divided between the visible and the near infra-red. Its effect upon the animal depends upon whether it is absorbed (and so adds to the heat burden to be dissipated) or is reflected. Red coat-colour reflects much more in the infra-red zone than does black. In reflectance, the Jersey is again at an advantage over most British breeds. Tests for suitability to a hot environment mostly involve determining the effect on body temperature, or respiration rate, of a standard extra heat load imposed by radiation or forced exercise. Provision of shade—with, as far as possible, good air movement—can improve the environment for cattle, even in England, on a hot summer's day.

Requirements for meat production

Meat consists primarily of muscle and fat. Though there can be too much fat, some is not only unavoidable but is also desirable. Even boned-out 'fat free' beef contains some 10 to 12 per cent of chemical fat. This must be remembered when interpreting the modern tendency to consider meat simply as an agglomeration of water, protein, ash and fat, of which only protein is important.

If a butcher buys cattle on a liveweight basis, it is naturally important to him that they should kill out well—that the saleable parts of the animal should be a high proportion of the whole. Similarly, the proportions of the carcase are important: not all parts are equally valuable, and his profit may depend upon the proportion of high- and low-priced joints (Fig. 4.36). Broadly speaking, the high-priced parts of the carcase are in the hindquarter and along the back; the underside, lower limbs and neck region provide the cheaper cuts.

When the British beef breeds were first developed, animals were fed to grow more slowly, and killed at much greater ages, and at heavier weights,

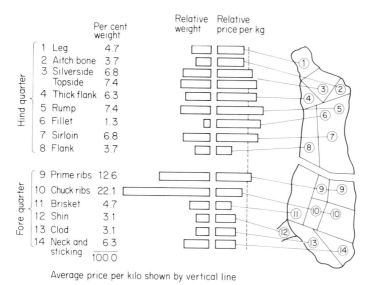

		Per cent weight	
Hind quarter	1 Leg	4.7	
	2 Aitch bone	3.7	
	3 Silverside	6.8	
	Topside	7.4	
	4 Thick flank	6.3	
	5 Rump	7.4	
	6 Fillet	1.3	
	7 Sirloin	6.8	
	8 Flank	3.7	
Fore quarter	9 Prime ribs	12.6	
	10 Chuck ribs	22.1	
	11 Brisket	4.7	
	12 Shin	3.1	
	13 Clod	3.1	
	14 Neck and sticking	6.3	
		100.0	

Average price per kilo shown by vertical line

Fig. 4.36 Comparative weights and prices of the different cuts in a beef carcase. (Modified from Short, J. B. (1928). *The Butcher's Shop.* Economics Research Institute, Oxford.)

than is now customary. In those days there was need for a higher energy intake in a man's diet (so much was done by human exertion that is now done by machines) and animal fat was a valuable article of diet. Since then, tropical vegetable oils have largely displaced animal fats, and fat has lost much of its importance as an accompaniment to muscle in meat.

Various factors, to different degrees in different circumstances, affect the value of meat. The most important are ease of cooking, flavour or palatability, tenderness, and size of joint. Meat that can be cooked by roasting or grilling is more valuable than that which requires stewing, for example. Meat from younger animals is generally more tender, and so the proportion of the animal that can be used for grilling is greater.

Joints for roasting are usually sold with the bone in; for them, the proportion of meat to bone does not matter, directly, to the butcher. However, the apparent proportions make a considerable difference to saleability (see Fig. 4.37). For conventional roasting, sufficient fat cover is required over the joint to prevent the meat from drying out during cooking, but excess fat has to be trimmed away, and is of very little value. The size of the carcase determines the size of a joint—or its shape at a given weight. Probably a decrease in size of families has been a considerable factor in the modern tendency to killing at lower carcase weights.

Most meat for grilling is sold off the bone, as also are some roasting joints. Here the proportion of bone to bone-free meat in a carcase is a matter that directly affects the butcher's profit, and thus determines the type of carcase that he seeks to buy.

The modern requirement is for young beef yielding a moderately small

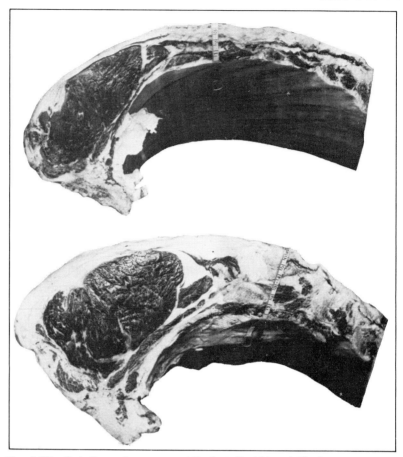

Fig. 4.37   Cuts through carcases at the last rib. (*Above*) Shallow eye-muscle
and large proportion of bone. (*Below*) Deep fleshed with a small proportion of
bone but with too much fat for the modern British market. (Hammond, J. and
Mansfield, W. S. (1936). *Journal of the Ministry of Agriculture, London*, **42**,
977.)

carcase (210–250 kg) without an excess of fat to be trimmed off. In the
U.S.A. there is a demand for beef with well-developed marbling fat (fat
laid down within the muscle, between the bundles of muscle fibres). To get
such meat there also has to be an excess of fat elsewhere, which has to be
trimmed off: however the U.S. market is willing to pay for this.

Development of beef conformation

The changes in body proportions that take place during growth are illustrated in Fig. 4.38 (see also Fig. 5.19). The calf with its relatively large
head, neck and legs has a high proportion of the low-priced parts of the

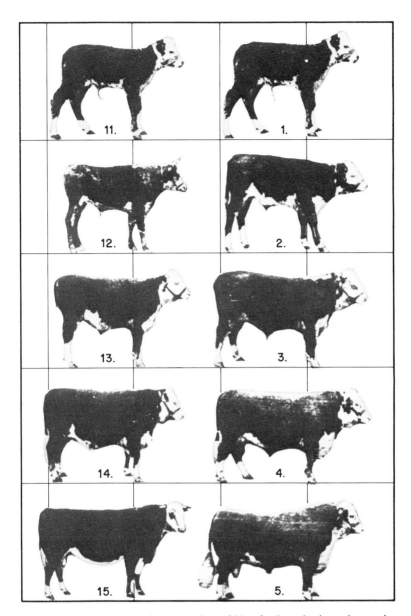

Fig. 4.38 Changes in body proportion of Hereford cattle. In order to show changes in proportions, as distinct from size, all the photographs are reduced to the same height at the shoulders.

11 Bull, 2 days

12 Steer, 30 months, grown on low
    level of nutrition

13 Steer, 11 months, grown on high
    level of nutrition

14 Steer, 22 months, grown on high
    level of nutrition

15 Bull of 100 years ago

1 Bull, 2 days

2 Bull, 5 weeks

3 Bull, 13 months

4 Bull, 22 months

5 Bull, 5 years

(Hammond, J. (1935). *Empire Journal of Experimental Agriculture*, **3(9)**, 1.)

Table 4.8 High-priced joints as a percentage of the total weight of joints, and muscle in high-priced joints as a percentage of the total weight of muscle. (From Pomeroy, R. W., Williams, D. R., Owers, A. C. and Scott, B. M. (1966). *A Comparison of the Growth of Different Types of Cattle for Beef Production.* Royal Smithfield Club, London.)

Stage of growth	Hereford		Hereford × Friesian		Friesian	
	Joints %	Muscle %	Joints %	Muscle %	Joints %	Muscle %
Calf	60.9	68.9	60.9	68.4	59.0	68.4
6 months	63.7	69.2	63.7	69.3	64.7	71.9
12 months	65.5	70.6	65.7	71.1	65.5	71.0
18 months	65.3	70.4	65.6	70.6	65.3	70.9
24 months	66.3	71.9	66.5	70.4	66.1	71.9

body. As it grows up, the body lengthens: the back grows at a faster rate than the low-priced joints (Table 4.8). The rate and extent to which these changes take place determine the value of the animal for beef purposes.

As growth continues, the body deepens; if a bullock of one of our improved beef breeds—which have been bred for killing at an early age ($1\frac{1}{2}$ to $2\frac{1}{2}$ years)—is kept on to heavy weights, this phase of development may go so far in depth of rib as to increase the proportion of this low-priced part, and so reduce the value of the carcase. In the male the lengthening and deepening, and extent of muscle development, are greater than in the female or castrate; but the male also shows greater development of the head and neck.

The proportions, and so the value, of the animal at a given age or weight are affected by the plane of nutrition on which the animal is reared. On a low plane of nutrition the earlier developing parts (head, heart and bones, for example) are less affected than later-developing parts and tissues. Animals reared to light market weights on a low plane of nutrition have a high proportion of low-priced joints, and their conformation appears similar to that of relatively unimproved types, even though well bred for beef (compare No. 12 with No. 14 in Fig. 4.38).

The beef breeds were developed from draught animals, selecting them for depth of fleshing and capacity for rapid growth; also for their ability to deposit fat, in particular fat deposited between the muscle bundles. Improvement for beef conformation has consisted partly in shortening the bones, and so thickening the muscle that lies over them. This gives a carcase with blocky joints and a great depth of flesh over them (Fig. 4.37). It has not in fact much altered the relative proportion of meat and bone. For the butcher, the advantage of the beef over the dairy type of animal is in the appearance of the joints; for the farmer the advantage is in speed of growth, and so in efficiency of food conversion.

In the improvement of beef breeds, animals have been reared on a high plane of nutrition, with selection for breeding of those which went through the age changes in proportion quickest and to the fullest extent. The adult Hereford bull of 1830, for example, is more like a 13-month-old modern bull than a present-day adult (compare Nos 15, 3 and 5 in Fig. 4.38).

Fig. 4.39 Champion Shorthorn steers; Smithfield, 1835, 1000 kg and Chicago, 1937, 500 kg. (Courtesy of R. G. Freer, New South Wales Department of Agriculture.)

Market requirements change, and the direction of breed improvement has to change to meet the altered needs. Formerly the emphasis was on early maturity to produce small blocky joints with adequate fat. Now less importance is attached to the appearance of joints, and more to rapid growth and the production of young lean meat. By analogy with the pig (see Fig. 6.12), one might expect this to lead to selection for the lengthening, rather than the deepening, phase of body growth.

In the United Kingdom the dairy herds are now the main source of calves for beef production, and the bull used to produce them is required to breed calves with a potential for rapid growth, without excess fat deposition before slaughter weight has been attained. Thus the former trend (Fig. 4.39) has been reversed and the need is for a large breed, such as the Charolais, with late maturity. Beef bulls are now being progeny tested on dairy cows (see Chapter 12).

Muscular growth and development

The muscle fibres in a muscle are tied together in bundles by connective tissue (mainly collagen); between these bundles, to an extent determined by breed, there may be cells which can lay down fat. The growth of the muscle which occurs after birth is largely by increase in the length and thickness of muscle fibres (Figs 2.4 and 4.40). In some muscles, such as the vastus externus the muscle fibres grow much larger than in others, such as the gracilis muscle. This increase in fibre diameter means increase in the size of the muscle bundles, and hence increased coarseness in the grain of the meat (Fig. 4.41).

Fig. 4.40 The effect of age on the diameter of muscle fibres of Suffolk rams. Sections through the fibres of the semi-membranosus muscle of a lamb at birth (*left*) and at 5 months (*right*). (Hammond, J. (1932). *Growth and the Development of Mutton Qualities in the Sheep.* Oliver & Boyd, Edinburgh.)

For this reason the meat of younger animals is finer grained than the meat of older ones. For this reason too some muscles, such as the psoas (undercut, fillet steak) are preferred to others, such as the rump muscles, of the same carcase. Small species and breeds of animals have smaller muscle bundles, and finer-grained meat than large ones, and so have been preferred for meat purposes. A large breed, however, killed at an early age, has fine-grained meat.

There are ageing changes in the collagen which ties the muscle fibres into

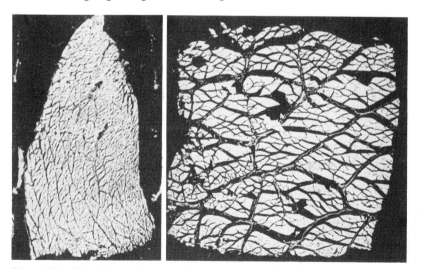

Fig. 4.41 The effect of age on the coarseness of grain in a muscle, the vastus externus, of Suffolk rams. (*Left*) At birth; (*right*) at 5 months old. (Hammond, J. (1932). *Growth and the Development of Mutton Qualities in the Sheep.* Oliver & Boyd, Edinburgh.)

bundles. In the young animal the collagen breaks down easily on cooking, to form gelatine. Proper hanging of the carcase aids this breakdown, but in the old animal breakdown is less easy or complete, and the meat from older animals can consequently be stringy, particularly in coarse grained muscles.

The muscles of the young animal are eaily fatigued, whereas as the animal grows up the power of sustained action increases. With this change in muscle comes an increase in the respiratory pigment, myoglobin, of the muscle and the colour of the muscle deepens. With increase in the colour in the muscle also comes increased flavour. Thus, for example, young veal is pale in colour and relatively flavourless when compared to beef, which is dark in colour. Beef, however, may be too dark and too highly flavoured to suit the public taste as is the case with old bull beef. Castration, lack of exercise, and lack of access to iron reduce the colour.

With the increase in muscle content of myoglobin (which provides a reserve of oxygen) there is also increase of the energy reserves within the muscle fibres, in the form of glycogen and of fatty material. Flavours are lipoid soluble substances, and the fat within the muscle is largely responsible for its flavour. Some flavour develops by bacterial action during the hanging of the meat, and some is transmitted from the bone during the cooking process.

Fat development

Just as the parts of the body develop in a definite order, so do the tissues. Fat, in general, is the last-developing tissue: it serves as an energy reserve, to be drawn on during periods of poor nutrition. Fat is not deposited uniformly throughout the body; there are early- and late-developing fat depots. In the early stages of fattening it is put on in the kidney fat; in those breeds unimproved for meat production, such as in the Jersey cow, fattening is not usually taken much beyond this stage.

Next, fat goes on in a subcutaneous layer, and so gives that smooth rounded appearance to the well-finished bullock. In the later stages of fattening, there is deposition of fat between the bundles of muscle fibres, as the 'marbling' fat, which is particularly valuable in the older bullock because it breaks up the muscle bundles and so makes the meat more tender. There is more marbling fat in those breeds which are early-maturing and fatten most easily. Palatability increases with the amount of fat in the meat, up to an optimum level which depends upon the individual; above the optimum level palatability decreases sharply.

When a bullock is fattened, proportionately more fat is added to the carcase than to the offal parts, and so the dressing-out percentage (or killing-out, or carcase, percentage) goes up. The carcase percentage (other things, such as breed and amount of stomach contents, being equal) is, therefore, a guide to the composition of the carcase. From Fig. 4.42 it can be seen that between carcase percentages of 50 and 60 there is a threefold increase in the percentage of fat and a halving of the proportion of bone.

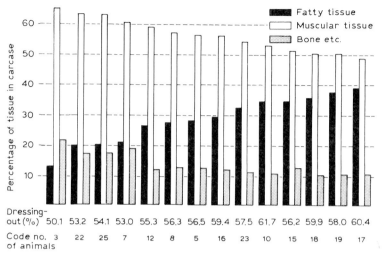

Fig. 4.42 Changes in the percentage of fatty tissue, muscular tissue, bone, etc. in the carcase of steers during fattening. (After Callow, E. H. (1944). *Journal of Agricultural Science*, **34**, 177.)

The carcase percentage therefore gives a good indication of the meat value of the animal. A high degree of finish is generally associated with a high degree of marbling, but is wasteful. Thickness of subcutaneous fat in the loin region provides a useful measure of carcase fatness (Wilson, 1967).

Table 4.9 gives killing-out percentage and carcase composition for buffalo fed for slaughter at various ages. Besides illustrating the greater fatness of the steer than of the entire animal, it also shows—by comparison with Fig. 4.42—the lower fat content of the buffalo. The initial fall with ageing in the dressing out percentage reflects the relatively late development of the ruminant stomach.

As an animal puts on fat it also stores within the fat a certain amount of yellow colouring matter (carotin). The pigment derives from the feed, from

Table 4.9 Dressing percentage and carcase composition of buffalo according to age and weight (Ragab, M. T., Darwish, M. Y. H. and Malek, A. G. A. (1966). *Journal of Animal Production of the United Arab Republic*, **6**, 9.)

Type of animal	Age at slaughter (months)	Average live weight (kg)	Dressing percentage	Percentage of carcase wt		
				Lean	Fat	Bone
Calf	1.7	74.0	57.4	68.5	6.3	24.9
	6	157.7	55.0	69.9	5.7	24.3
Bull	12	230.3	53.7	67.0	11.8	20.5
	18	359.3	55.8	67.5	13.5	18.8
	24	449.0	52.7	69.8	12.6	16.3
Steer	12	236.0	53.3	64.8	15.1	18.9
	18	360.3	52.8	66.8	13.9	18.5
	24	450.4	54.3	66.5	15.0	17.3

plant pigments mainly found in the green parts of the plant. The amount of pigment stored depends upon the breed as well as upon the diet. If an animal goes back in condition, and uses up its fat, the pigment becomes concentrated. Thus the fat of an old animal such as a dairy cow, tends to be deeply coloured. Deep yellow fat is not liked by the consumer. The flesh of 'barley-beef' animals is easily recognized by the pale colour and firm texture of the fat. The hardness of the fat depends upon its position in the body and upon the food from which it is derived. Thus subcutaneous fat is softer than kidney fat, and fat laid down when fattening on grass (which contains unsaturated fats) is softer than the saturated fats which the barley-beef animal manufactures from starch.

Growth in liveweight

The rate at which an animal will grow, fatten or give milk is limited largely by the food intake per day. In stall feeding this can be controlled by the preparation of the food and by the use of feeds low in fibre and high in energy and protein. In grazing pasture, however, other factors come into play. As shown by Johnston-Wallace and Kennedy, cattle graze only for about 8 hours in a day and, having only a 10 cm span of incisor teeth and pad, their food intake is limited by the density and length of the herbage. Thus they find that when the herbage is 10 to 13 cm high the maximum amount of food, sufficient for the production of 24 kg of milk per day (70 kg), can be eaten. However when the herbage reaches 25 cm high it is less dense, so that only 35 kg can be eaten. In good fattening pastures the herbage is very dense, so that a large mouthful is obtained at every bite.

Store periods, whether they be in the winter as in the United Kingdom or the summer as in Australia, cause a great loss in feed (see Fig. 8.3). The conversion of feed stuffs into animal products entails losses (see Table 4.10); efficiency of conversion is increased by increasing growth rate, or

Table 4.10   The efficiency with which feeding-stuffs are converted into animal products by different types of livestock. (From Halnan, E. T. (1944). *Proceedings of the Nutrition Society*, **1**, 32.)

Product	Protein conversion efficiency %	Energy conversion efficiency %
Milk (3 lactations)	17.0	30.0
Eggs (2 years)	33.1	22.1
Poultry meat–cockerels (1.6 kg)	17.9	15.6
pullets (1.6 kg)	18.2	12.5
Beef (birth to 360 kg)	11.1	14.0
,, (birth to 460 kg)	8.8	19.0
,, (grass fattened, birth to 540 kg)	7.3	15.1
,, (stall fattened, birth to 640 kg)	7.2	15.3
Pork	13.8 to 16.0	34.4 to 39.4
Bacon	12.4 to 13.8	36.4 to 43.4
Lamb	5.9	9.6

level of production. It takes about three units of energy in feeding stuffs to produce one unit of energy as fat in beef, but when the animal has to use this fat for maintenance during the store period it is equivalent only to one unit of energy in oats or other feeding stuffs.

Meat is more expensive than cereals for human food: and for animal food also. When a young animal loses weight during a store period it takes off fat and muscle, but its skeleton continues to grow. When eventually it reaches marketing weight its body proportions may differ from those of an animal grown with no store period.

Whether or not it is profitable to avoid a store period depends upon local conditions. Continuous growth and early marketing in the U.S.A. are achieved under feed-lot conditions, and in the U.K. by the 'barley-beef' system.

As the animal grows the proportion of fat in its carcase increases; when it grows more slowly it will reach a greater weight before it attains the same proportion of fat. The age for slaughter at optimum degree of fatness will vary with speed of growth and with breed capacity to lay down fat. For the British market, for steers grown continuously from birth, the ages found by Pomeroy *et al.* in 1966 were 12 to 13 months for Herefords, 15 months for Hereford–Friesian crosses and 17 to 18 months for Friesians. The Friesian–Charolais cross fattens less readily than the Friesian–Hereford (Table 4.11).

The influence of breed and sex is well illustrated (Fig. 4.43) by a survey of the relation between breed, sex, liveweight and price of cattle sold in East Anglian markets. The average liveweights were 500 kg for steers and 450 kg for heifers. For weights above these, prices fall more steeply for the Shorthorns (which carry more fat) than for the Friesians. The heifer has a smaller frame than the steer, and reaches a corresponding degree of fatness at a lower body weight.

Growth rate in steers on a high plane of nutrition, and also in wethers (but not in pigs), can be increased by the administration of synthetic oestrogen—either by addition to the feed, or by subcutaneous implantation of a tablet in the ear. Such treatment has been widely employed in the

Table 4.11 Comparison of bone, fat and muscle in the carcases of Charolais × Friesian and Hereford × Friesian cattle. (From Edwards, J., Jobst, D., Hodges, J., Leyburn, M., O'Connor, L. K., McDonald, A., Smith, G. F. and Wood, P. (1966). *The Charolais Report*. Milk Marketing Board, Thames Ditton.)

Tissue	Percentage tissue in whole side			
	A. Based on total weight		B. Based on fat free weight	
	Charolais × Friesian	Hereford × Friesian	Charolais × Friesian	Hereford × Friesian
Bone	14.6	13.8	18.6	19.6
Fat	21.6	29.5	–	–
Muscle	59.3	52.3	75.6	74.2
Remainder	4.5	4.4	5.7	6.2

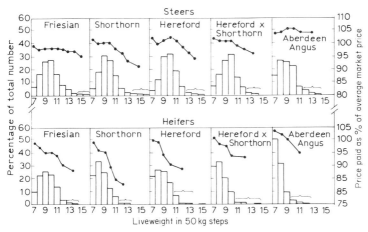

Fig. 4.43 Distribution of live weights of steers and heifers sold in East Anglian markets in 1956–7, together with the relative prices paid per kg for cattle of different weights. 7 on scale includes cattle from 350–400 kg, etc. (Everitt, G. C. (1958). *A Survey of Beef Production in the East Anglian Area of England*. School of Agriculture, Cambridge.)

U.S.A. There may be a stimulation of appetite, but the principal effect is an alteration in the pattern of growth, more muscle and less fat being laid down. Consequently growth rate is increased (see page 33). The effect is illustrated in Table 4.12.

Table 4.12 Effect of stilboestrol implantation on fattening cattle. (From Lamming, G. E. (1958). *Journal of the Royal Agricultural Society of England*, **119**, 41.)

Trial	No. of animals	Dose (mg)	Time and feeding	Average daily gain (kg)		Percent increase Treated/ Control
				Treated	Control	
1	17 steers	60	Winter: yard	0.91	0.55	63
2	23 steers	36	Spring: pasture	1.24	1.10	15
3	62 steers	24	Summer: pasture	0.73	0.55	33
		36	Summer: pasture	0.86	0.56	53
4	20 steers	36	Autumn: pasture	1.18	0.77	55
	21 heifers	36	Autumn: pasture	0.95	0.82	15

Little benefit is to be expected from treatment of animals in early stages of growth or on a low plane of nutrition, for under these conditions there would be little fat deposition in untreated animals. The greater benefit noted in autumn and winter (Table 4.12) suggests a greater tendency to deposit fat at these seasons, as with many wild species.

In many countries of Europe young bulls of about 12 months old are now being killed for beef. They utilize food more efficiently than do steers, do not over-fatten, and grow more rapidly (Table 4.13). With older bulls management is a problem, as also is the darker flesh and associated stronger flavour. These difficulties are not encountered with young bulls.

Table 4.13 Comparison of growth rate and carcases of bulls and steers. (From Prescott, J. H. D. and Lamming, G. E. (1964). *Journal of Agricultural Science*, **63**, 341.)

Character	Bulls	Steers
Daily gain in weight (kg)	1.05	0.91
Weight at slaughter (kg)	405	378
Composition of 10th rib (%)		
Bone	19.0	19.0
Fat	16.8	29.2
Muscle	64.2	51.8

There is the prospect that treatment with synthetic anabolic substances (p. 13) might stimulate muscle growth in steers without causing the management problems posed by natural testis hormone production. However (as with stilboestrol) the analogues available for such treatment owe their potency to their being less readily inactivated than the natural hormones. Many years ago women were sometimes (most misguidedly) treated in pregnancy with large doses of stilboestrol; many years later some of their daughters—presumably exposed at a critical prenatal stage of evolution—developed a cancer of the vagina. This has resulted in a somewhat natural public prejudice against eating meat of animals treated with sex hormone analogues—no matter how small the residue of the drug in the meat.

References

BICKERSTAFFE, R., ANNISON, E. F. and LINZELL, J. L. (1974). The metabolism of glucose, acetate, lipids and amino acids in lactating dairy cows. *Journal of Agricultural Science*, **82**, 71.

BILTON, R. J. and MOORE, N. W. (1977). Successful transport of frozen cattle embryos from New Zealand to Australia. *Journal of Reproduction and Fertility*, **50**, 363.

CASSOU, R. (1968). La miniaturization des paillettes. *Proceedings of the 6th Congrès International de Reproduction et Insémination Artificielle, Paris*, **2**, 1009.

HAFEZ, E. S. E. (1955). Puberty in the buffalo-cow. *Journal of Agricultural Science*, **46**, 137.

HAMMOND, J. (1936). The physiology of milk and butterfat secretion. *Veterinary Record*, **16**, 519.

HANSEL, W. and TRIMBERGER, G. W. (1951). Atropine blockage of ovulation in the cow and its possible significance. *Journal of Animal Science*, **10**, 719.

HEAP, R. B., HOLDSWORTH, R. J., GADSBY, J. E., LAING, J. A. and WALTERS, D. E. (1976). Pregnancy diagnosis in the cow from milk progesterone concentration. *British Veterinary Journal*, **132**, 445.

MEHTA, S. N., GANGWAR, P. C., SHRIVASTALA, R. K. and DHINGRA, D. P. (1979). Effect of cooling on reproductive behaviour in buffaloes. *Journal of Agricultural Science*, **93**, 249.

NAGASE, H. and NIWA, T. (1964). Deep freezing bull semen in concentrated pellet semen. I, II, III. *Proceedings of the 5th Congresso Internationale per la Riproduzione Animale e la Fecondazione Artificiale, Trento*, **4**, 410, 498, 502.

ROY, J. H. B., GILLIES, C. M., JOHNSON, V. W., GANDERTON, P., STOBO, I. J. F. and POPE, G. S. (1977–78). Early breeding of dairy heifers. *National Institute for Research in Dairying Biennial Report, Reading*, p. 127.

SAUMANDE, J. (1978). Relationship between ovarian stimulation by PMSG and steroid secretion. In *Control of Reproduction in the Cow* (J. M. Sreenan, Ed.). p. 169. Martinus Nijkoff, The Hague.

SMITH, K. L. and SHAUBACHER, F. L. (1973). Hormone induced lactation in the bovine. I. Lactational performance following injections of β-oestradiol and progesterone. *Journal of Dairy Science*, **56**, 738.

TAYLOR, ST. C. S., MONTEIRO, L. S., MURRAY, J. and OSMOND, T. J. (1978). Possible association between the breeding value of dairy bulls and milk yields of their mates. *Animal Production*, **27**, 303.

THIBAULT, C., GERARD, M. and MENEZO, Y. (1975). Acquistion par l'ovocyte de lapine et de veau du facteur de décondensation du noyau du spermatozoïde fecondant (MPGF). *Annales de Biologie Animale Biochemie Biophysique*, **15**, 705.

WEBSTER, A. J. F. (1976). The influence of the climatic environment on metabolism in cattle. In *Principles of Cattle Nutrition* (H. Swann and W. H. Broster, Eds), p. 103. Butterworth, London.

WELCH, R. A. S., CRAWFORD, J. E. and DUNGANZICH, D. M. (1977). Induced parturition with corticoids; a comparison of four treatments. *New Zealand Veterinary Journal*, **25**, 111.

WILSON, P. N. (1967). The relationship of the beef animal to the final meat product. *Bulletin of the Institute of Meat*, No. 57 (August 1967), 28.

WOOD, P. D. P. (1975). A note on the effect of twin births on production in the subsequent lactation. *Animal Production*, **20**, 421.

WRIGHT, N. C. (1954). The ecology of domesticated animals. In *Progress in the Physiology of Farm Animals* (J. Hammond, Ed.), p. 191. Butterworth, London.

Further reading

BERG, R. T. and BUTTERFIELD, R. M. (1976). *New Concepts of Cattle Growth*. Sydney University Press.

BETTERIDGE, K. J. (Ed.) (1977). *Embryo Transfer in Farm Animals – A Review of Techniques and Applications*. Monograph No. 16, Canadian Department of Agriculture.

BROSTER, W. H. (1972). Effect on milk yield of the cow of the level of feeding during lactation. *Dairy Science Abstracts*, **34**, 265.

HAMMOND, J. (1932). *Growth and Development of the Mutton Qualities in the Sheep*. Oliver and Boyd, Edinburgh.

M.A.A.F. (1975). *Energy Allowances and Feeding Systems for Ruminants*. Ministry of Agriculture Fisheries and Food Technical Bulletin, No. 33. H.M.S.O., London.

SREENAN, J. M. (Ed.) (1978). *Control of Reproduction in the Cow*. E.E.C. Seminar, Galway. Martinus Nijkoff, The Hague.

WILLIAMSON, G. and PAYNE, W. J. A. (1978). *Animal Husbandry in the Tropics*, third edition. Longman, London and New York.

5 Sheep and goats

The breeding season

The period of the year at which oestrous cycles occur has been evolved by natural selection so that the young are born at the time of year that will give them their best chance of survival. Those breeds which have originated in northern latitudes such as Iceland and Scotland have, in general, a short breeding season, for if the lambs are born too early in the year they will perish from cold, whereas if born too late they will not be sufficiently well grown to survive the following winter. Breeds which have originated nearer the equator, such as the Merino (from Spain) have a much more extended breeding season. Those breeds which have originated in areas where the winter is cold and prolonged, northerly (Blackface) or at a high altitude (Welsh), have fewer heat periods in the breeding season than those which have originated in more southerly (Suffolk) or lowland (Romney) areas—see Fig. 5.1. The Welsh has, on the average, only 7 heats in the breeding season while the Dorset Horn has 13 and their crosses have 10.

The time of onset, and duration, of the breeding season have been shown by Yeates (1949) to be affected by day length. Yeates did this by housing sheep from late afternoon until the following morning and by controlling the day-length for his experimental group by providing artificial light (when required) in an otherwise darkened building. His control sheep were in a pen with windows. By altering the pattern of day-length change he, in effect, at the beginning of the breeding season transported them across the equator—as if to Australia or New Zealand—though of course the climatic conditions (other than of day-length) remained those of Cambridge, England. In the experimental ewes the breeding season ended early, and their following season began at much the same time as it would have done in similar sheep kept in Australia. Goat does respond similarly.

In most of those breeds in which the extent of the breeding season has been measured, it is fairly evenly distributed either side of the shortest day; but this is not so in the Dorset Horn (Fig. 5.1) which may breed at midsummer, nor of some Merinos in which the season can begin while day-length is still increasing. In lambs, the age at first heat is determined by an interaction between age, day-length, and nutritional status. The breeding season of lambs starts later than that of adult ewes: late-born lambs are not on heat so soon, yet they come on heat at an earlier age;

while lambs born very late do not breed until the season of the year following.

When breeding from ewe lambs it is normally advisable to put them to the ram about 6 weeks later than the older ewes, because their breeding season begins later, and also because nutritive conditions will be better for them during the latter part of pregnancy. For the practice to be worthwhile it is necessary that the lambs mated should be well grown and able to be well fed during pregnancy. It is wise, also, to join maidens separately from mature ewes, as the latter exhibit more aggressive oestrous behaviour and may monopolize the rams (Lindsay, 1966).

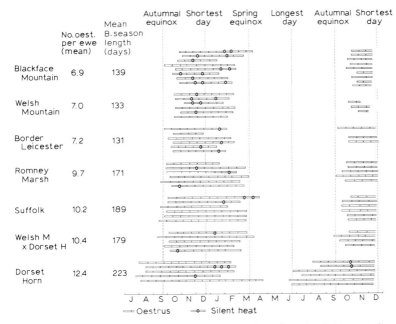

Fig. 5.1   Duration of the breeding season in various breeds of sheep at Cambridge in relation to daylight hours. Breeds originating in high latitudes (Blackface) or altitudes (Welsh) have a later and shorter breeding season than those from low latitudes (Romney) or altitudes (Suffolk), while crosses between the two (Welsh × Dorset Horn) are intermediate between the parent breeds. (Hafez, E. S. E. (1952). *Journal of Agricultural Science*, **42**, 189.)

Most British breeds do not naturally lamb until after the breeding season has ended. The Dorset Horn and the Merino can be mated naturally to lamb within the season. The majority of ewes lambing in the first part of the breeding season come on heat 20 to 60 days later: if lambs are weaned, or die, the ewes come into heat a little earlier. As in the cow, fertility to service soon after parturition is low.

The first ovulation of the breeding season is 'silent'—it is not accompanied by oestrus—and the ewe ovulates spontaneously. If a ram is introduced into a flock of ewes shortly before the first ovulation of the season is due, the timing of that ovulation, as evidenced by the number of ewes coming on heat about 3 weeks later, is affected by the introduction of the ram. The smell of the sebaceous and sweat gland secretions of the ram provides the main stimulus (Knight and Lynch, 1980). Probably the odour of the male has a similar function in the goat; it is well known to increase with the approach of the breeding season.

The length of the oestrous cycle of the sheep is generally 16 or 17 days, and normally within the range of 14 to 19 days; longer intervals are mainly due to the intervention of silent heats. Short cycles of 6 or 7 days are common in the goat at the beginning of the season, before regular cycles of 20 or 21 days commence. The measured duration of heat depends on the sex drive of both ewe and ram, and upon how closely they are confined together (Lindsay, 1966). The normal duration is usually considered to be 24 to 48 hours, and in such heats ovulation occurs near the end of the heat period. The time relationship between oestrus and ovulation when oestrus is very short (3 hours) or very long (up to 84 hours) is not known.

Fertility and sterility

Sterility due to anatomical deformity is rare, but temporary infertility or even permanent sterility occurs in some parts of the world, due to the ingestion of plant oestrogens. This is particularly common in Australia, where clovers form a large part of the diet. The oestrogenic substances in clover are metabolized differently in the ruminant stomachs of sheep and of cattle, and cattle are not markedly affected.

Fertility, as measured by lambing performance, is a breed characteristic (Table 5.1) and can be altered by selective breeding. It can also be modified by age, stage of the breeding season when bred and by nutrition. The lambing percentage is primarily limited by the number of ova shed. Some breeds, such as the Australian Merino, commonly shed only one egg. Others, such as the Border Leicester, usually shed two, and the Romanov three or four eggs. Failure of fertilization is not generally considered to be a major factor in determining lambing percentage in adult ewes. However, the ewe does limit the number of eggs which implant (see Fig. 5.4), and in ewe lambs of breeds such as the Suffolk, twin ovulations are common, while twin births are few, and it is usual for a fair proportion of those mated to be barren. Fertility of matings at the height of the season, when

Table 5.1   Breed differences in lambing percentage. (From Asdell, S. A. (1964). *Patterns of Mammalian Reproduction*, 2nd edition. Constable, London.)

Breed	Lambs per 100 ewes	Breed	Lambs per 100 ewes
Cheviot	89	Dorset Horn	137
Scottish Blackface	93	Suffolk	144
Karakul	110	Shropshire	162
Corriedale	114	Border Leicester	181
Southdown	119	East Friesian	205
Romney Marsh	129	Romanov	238

oestrus is most intense (October, November in Britain) tend to give higher twinning percentages than matings earlier or later.

It has long been known that fertility is influenced by nutrition. It is commonly accepted, though not scientifically well authenticated, that fatness is a factor in for low fertility. The practice of 'flushing' (raising the plane of nutrition of the ewes before mating) is a long-established method for increasing ovulation and lambing rates. Only fairly recently has it been realized that the body condition at the time of mating, rather than a change in the level of feeding, is what primarily determines the ovulation rate. This is not to say that it is bad practice to keep the ewe flock in poorer condition during the summer and then to flush them before mating, and some experimental results indeed suggest there can be an additional beneficial effect from flushing on ovulation rate. Reduction in the level of feeding after mating has been investigated by Coop and Clark, who found no effect on reproductive performance and concluded that it is a useful practice because it results in a saving of feed which can better be supplied later in pregnancy.

Artificial control of breeding

If a ewe produces a single lamb once a year, the cost of that lamb at birth approximates the cost of keep of the ewe for a year, less the value of the fleece that it produces. If availability of feed allows, it is desirable that lambs should not be singles, and that ewes should breed more often than once a year. With milch goats which have a short breeding season and lactation, and which kid in the spring, there is the problem of maintaining an even year-round level of milk production, and with sheep one of providing lambs at all times of year. Maximum use of a ram or buck requires artificial insemination, and there is a saving of the time of the inseminator if all females can be inseminated together. Further, there can be advantages in having the whole of a flock or herd giving birth over a brief period—which can be expected if the time of ovulation is synchronized. Ovum transplantation offers possible benefits similar to those outlined for cattle (p. 69).

Mating outside the breeding season

An injection of pregnant mare serum hormone (MSG) will cause growth of follicles in an anoestrous ewe, and this is followed by ovulation. However, as at the start of the breeding season, the ewe does not come on heat. A second injection 16 days later commonly causes heat and a second ovulation, but fertility has not generally been good.

Treatment with progesterone or a suitable analogue for several days before the injection of MSG makes the ewe sensitive to the oestrogen from the ripening follicle, so that she will come on heat. Progestagen can be given by a subcutaneous implant, but intravaginal sponges (Fig. 5.2) are more convenient and more widely used. The sponge acts as an artificial

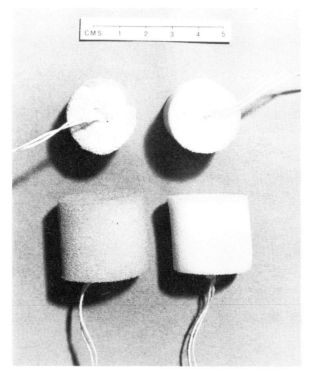

Fig. 5.2 Polyurethane sponges used for intravaginal or subcutaneous applications. The top left-hand sponge is impregnated with 800 mg progesterone (clearly visible) and the top right-hand sponge with 50 mg SC-9880 (not visible). The lower sponges are not impregnated. (Robinson, T. J. (1965). *Nature, London,* **206,** 39.)

corpus luteum, the progestagen being continually absorbed through the vaginal wall.

A suitable dose of MSG (400–600 i.u.) is injected when the sponge is removed and oestrus with ovulation occurs about 36–48 hours later. In the anoestrous goat, the MSG is injected 48 hours before sponge withdrawal.

In 1978, in France, some one million sheep and goats were treated in this way. As Table 5.2 shows, oestrus is consistently induced, but conception rate tends to be low in mid-anoestrus. The average number of lambs per pregnancy in these experiments was similar (1.6 or 1.7) at all three times of year. In late anoestrus, conception rate is fairly good, and ewes which fail to conceive return to service; but in mid-anoestrus conception is poor and the ewes fail to recycle (Table 5.2).

The poor conception rates seem to be due to poor sperm transport or survival in the female tract. Similar low fertility has been found in treated anoestrous goats with a high level of milk production. In such animals Corteel (1975) finds that insemination of much higher than usual numbers of sperm overcomes the otherwise poor fertility. Because treated animals

Table 5.2 Data on sponge-MSG treatment of ewes in Ireland, 1968–75. (From Gordon, I. (1977). *Symposium on Management of Reproduction in Sheep and Goats.* Madison, Wisconsin.)

	Season		
	Spring mid-anoestrus	Summer late-anoestrus	Autumn breeding season
No. treated	2508	21 545	1600
Per cent oestrous	93	97	97
Per cent pregnant			
First oestrus	35	64	75
First and second oestrus	35	80	91

are all on heat at much the same time, either artificial insemination, or else a high male to female ratio at joining, are necessary.

Synchronization of oestrus

In the breeding season the timing of oestrus can be synchronized by the same methods as apply to cattle (p. 66); either two treatments with prostaglandin can be given, or the cycle length can be prolonged by progestagen treatment. Impregnated vaginal sponges are withdrawn after 12–14 days in the sheep, or 18–20 days in the goat. The majority of animals come on heat two days later (Fig. 5.3). Injection of MSG (400 i.u.) at the time of

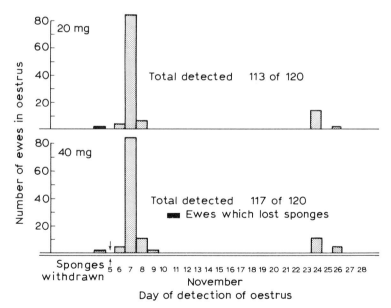

Fig. 5.3 The time of onset of oestrus after withdrawal of intravaginal sponges impregnated with 2 dose-levels of a highly active progestagen. (Robinson, T. J., Moore, N. W., Holst, P. J. and Smith, J. F. (1967). In *The Control of the Ovarian Cycle in the Sheep.* (T. J. Robinson, Ed.) Sydney University Press.)

sponge withdrawal induces a mild degree of superovulation and an earlier onset of heat (Evans and Robinson, 1980). The degree of synchronization of ovulation is such as to allow routine insemination without testing for oestrus.

Increasing the ovulation rate

As in cattle (page 68) the number of ovulations at natural oestrus can be increased by injecting MSG 3 or 4 days before the end of the cycle. With a dose of 500–750 i.u., 2 to 9 ova may be expected, and they will generally be fertilized. However there is early embryonic death of some ova (Fig. 5.4), and the capacity of the uterus of the particular ewe to support the

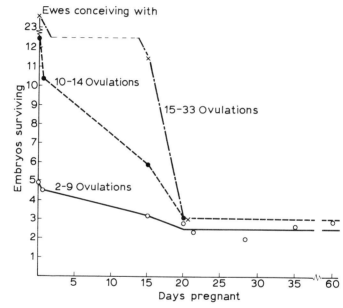

Fig. 5.4   Diagram showing how, no matter how many embryos are conceived as a result of injecting MSG before service, the number is reduced by embryonic death, mainly before the 20th day of pregnancy, to the maximum (on the average 2 to 3) that the individual ewe is capable of carrying. (Robinson, T. J. (1951). *Journal of Agricultural Science*, **41**, 6.)

eggs, rather than the number of eggs shed, would appear in practice to be what limits the increase obtainable. Viable quadruplet and quintuplet lambs have been produced occasionally. Experimentally, increases of lambing by 20 or 30 per cent have been obtained.

It is of course necessary to know when oestrus is due in order to inject the MSG at the correct time. These trials involved running the ewe flocks with raddled vasectomized rams, and identifying the individual ewes to be treated (each on the 12th or 13th day after it had been observed on heat).

More recently the necessity for this procedure has been avoided by synchronizing the whole flock (see above), injecting them all on the same day,
towards the end of the period of progestagen treatment, and later inseminating artificially.

Egg transplantation

Fertilized eggs produced by these techniques may be surgically transferred, two at a time, into recipient ewes which are at the same stage of the
cycle. The fertilized eggs can be kept alive outside the body in sheep blood
serum at 10 °C for 3 days and in the Fallopian tube of the rabbit for 5 days.

Fig. 5.5 Lambs born in South Africa following their importation from Great
Britain as fertilized ova. The ova were transported from Cambridge to Pieter-
maritzburg in the ligated oviducts of live rabbits and transferred on arrival to the
uteri of local ewes, shown here with the lambs. The 5-month-old Border
Leicester ram lambs (in the centre) with the Dorper ewes that reared them.
(Hunter, G. L., Bishop, G. P., Adams, C. E. and Rowson, L. E. A. (1962).
Journal of Reproduction and Fertility, **3**, 33.)

Fertilized eggs have been flown from Cambridge to South Africa in rabbits,
and live lambs born to foster mothers in that country (Fig. 5.5). The
technique has been used to increase the number of stud Poll Merino sheep,
and of Angora goats, in Australia. Nowadays, embryos can be frozen
before transfer, so that the need for synchronizing the cycles of donor and
recipient ewe or doe is eliminated.

The ram and artificial insemination

In many species with a restricted breeding season, the testes during the
non-breeding season revert to an immature condition. This is not true of
most breeds of sheep and goats but there is diminished sex drive and poorer

quality of ejaculated semen. Lack of drive in the ram has affected some experiments aimed at getting females to breed out of season.

Sterility is occasionally caused by infection of the epididymis. This, being a single much-coiled tubule, will be entirely blocked by quite a small lesion—with dilatation of the epididymis below the lesion. A physiological type of (temporary) sterility is caused by high environmental temperatures. It is the function of the scrotum to keep the testis at below body temperature. Those breeds with much wool on the scrotum are liable to become temporarily sterile in hot climates if the scrotum is not kept clipped. Exposure to 40.5 °C for only a few hours impairs fertility for 5 or 6 weeks, the approximate length of the cycle for production of new sperm.

Two methods are commonly used for collecting semen from the ram—with the artificial vagina or by inducing ejaculation by electrodes inserted into the rectum. These are described by Emmens and Robinson (see Fig. 5.6). The artificial vagina is similar to that used for the cow, but smaller, while the electro-ejaculator has been considerably modified and improved since Gunn developed the first model in 1936. A model developed at the Ruakura Animal Experiment Station in New Zealand (Fig. 5.6) is simple to use and does not cause violent spasm as did earlier models. Only one

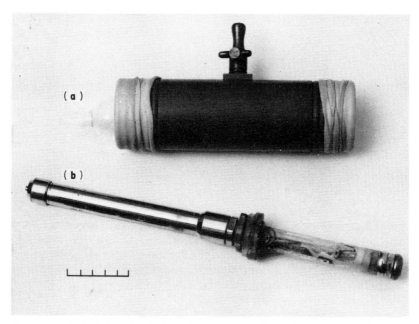

Fig. 5.6 Artificial vagina (a) and electro-ejaculator (b) used for collecting semen from rams. The electro-ejaculator is self-contained and is operated by torch batteries which fit into the metal handle. Ejaculation is caused by high-frequency electrical impulses between the two electrodes at the end of the perspex probe. The scale is in cm. (Emmens, C. W. and Robinson, T. J. (1962). In *The Semen of Animals and Artificial Insemination*. (J. P. Maule, Ed.) Commonwealth Agricultural Bureaux, Farnham Royal.)

collection a day can be made over prolonged periods, compared with 3 or 4 a day with the artificial vagina, so that the latter method is preferred, except for old or unfit rams.

Rams may readily be trained to serve into the artificial vagina. About 1 ml is usually produced, and up to 11 collections per day can be made for short periods. The semen is concentrated—about 3 to 5 thousand million sperm per ml. For artificial insemination, about 125 million are required to give maximum fertility; about 25 to 40 ewes can be inseminated with a single ejaculate.

When oestrus is synchronized (p. 115) in the breeding season twice this number are needed, and 500 million are required for out-of-season breeding (Colas, 1975). Semen is usually diluted about 1:1 or 1:2 with egg yolk citrate or with heated milk. For insemination the ewe is restrained, the vagina is opened with a speculum, and the cervix is located with a head light. About 0.1 ml of the diluted semen is deposited in the first fold of the cervix. A common method of restraint is to lift the hind quarters over a rail. Insemination of goat does is simpler because semen can be deposited directly into the uterus; consequently fewer spermatozoa are necessary.

Storage of ram and goat semen is not so well developed as for cattle. Semen cooled to 5 °C loses fertility rapidly after more than 24 h storage; though fertility to normal insemination drops to about 20 per cent after 3 days, fertilizing capacity is retained for up to 8 days if the semen is inseminated surgically into the Fallopian tube (Salamon, Maxwell and Firth, 1979).

Semen may be frozen and stored in liquid nitrogen by either the pellet method (Salamon, 1976) or in straws (Corteel, 1974; Colas, 1975); but semen from some rams and bucks will not freeze satisfactorily. Present methods require a degree of dilution before freezing such that the thawed semen must afterwards be concentrated by centrifugation in order to get a sufficient number of sperm into the small volume which is inseminated.

Diagnosis of pregnancy

Of the methods available in cattle (p. 76), the most direct—palpation of the uterus—is not available. If the date of mating is known, the milk progesterone test is applicable to the milking goat (or sheep). Measurement of blood progesterone level at the corresponding time should be equally reliable, but is of course not very convenient in practice.

In later pregnancy the blood concentration of placental lactogenic hormone is much greater in sheep and goats than it is in cattle; in goats the level of maternal pituitary lactogenic hormone is only about a tenth of the total activity present (Buttle et al., 1979). Measurement of this hormone is thus a possible, though not convenient, method of diagnosis. A field test, not applicable until after about day 60, employs echo sounding (as in the measurement of pig back-fat thickness) to detect the presence of fluid in the uterus. Preliminary results (Trapp and Slyter, 1979) indicate that it is about 90 per cent accurate.

Climatic adaptation

The lower critical temperature for closely shorn sheep—even when on a
high plane of nutrition—is above 20 °C (see Fig. 2.1). However the fleece
provides such a degree of insulation as to permit survival over winter even
on the hills of Scotland. Such mountain breeds have a coarse outer coat
which serves to drain off rain and so keep dry the fine undercoat, which
insulates by retaining a layer of still air. The fine wooled Merino breed has
no such outer coat; its dense fleece absorbs rain and dries out slowly.
Evaporation from the wet fleece chills the animal, and bacteria in the damp
wool generate odours which attract blowflies. The young Merino lamb is
particularly vulnerable to cold and wet. However in a dry climate the
Merino is most effectively insulated against both cold and heat.

Though the sheep is well provided with sweat glands, it seems that these
can have relatively little effect in wool sheep in regulating heat loss. The
main means of temperature regulation is by controlling evaporative loss

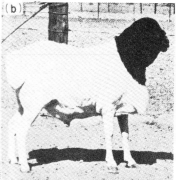

Fig. 5.7 Adaptive mechanisms to heat in sheep and goats. **(a)** Indian Loli
ram; short wool coat and long ears. **(b)** Blackheaded Persian ram; fat stores on
the brisket and in the tail. **(c)** Sudanese Desert sheep; short smooth coat and
long ears. **(d)** Jumnapari goat buck; long open coat and very long ears. (From
Williamson, G. and Payne, W. J. A. (1959). *Animal Husbandry in the Tropics.*
Longmans, London.)

through the lungs; rapid shallow respiration entails no greater loss of carbon dioxide (and consequent disturbance of blood hydrogen ion concentration) than does slow deep breathing—but it allows greater loss of water vapour.

Breeds of sheep (and goats) adapted to hot climates show the same general sorts of anatomical modifications as are seen in cattle (p. 94). Apart from a short wool coat, or the replacement of wool by hair, there is (Fig. 5.7) the localization of subcutaneous fat to one or two areas, such as the rump or tail, and increased body surface area for heat loss—such as the large pendulous ears of the Nubian goat.

Birth weight

The birth weight of the lamb or kid is affected by the number at birth, the level of nutrition, its sex, and the age and breed of the dam. During the early stages of pregnancy, however, little or no difference in weight exists, but the size and strength of the lamb at birth can be considerably affected by the state of nutrition of the ewe during the second half of pregnancy. Vérges found (Fig. 5.8) that when ewes were fed so as to gain 18 kg during the last 53 days of pregnancy, as compared with ewes which were fed so as only to gain 0.5 kg during that period, the twin lambs were 47 per cent larger at birth and weighed 4.1 kg as compared with 2.8 kg. The weight of the single lambs, however, was but little affected, as the nutrients required for a single lamb could easily be supplied from the mother's own body. Every year many lambs are lost or severely checked in their growth by insufficient nutrition of the ewes during the latter part of pregnancy.

As Fig. 5.8 shows, not only were twins from the high-plane ewes larger, they were also more mature in their body composition and had much more fat. The fat is important to the lamb in two ways: as subcutaneous insulation and, particularly in 'brown fat', as a reservoir of energy. Brown fat is a special form of adipose tissue which, in emergency, serves as a reserve source of body heat (particularly in hibernating animals and in the newborn). The less mature lamb, or one already exposed to stress in a prolonged birth, has less energy reserves from which to increase its heat production to balance losses. As well as being more mature in composition, the heavier lamb is physiologically more mature in capacity to maintain its body temperature.

Extra food supplied during the second half of pregnancy not only causes increased growth of the lamb but also increases the development of the udder, so that more milk is secreted after parturition and the lambs grow much quicker after birth. For example, twin ram lambs have been found to weigh 30 kg at 13 weeks old where the ewes were well fed during the second half of pregnancy, as compared with only 18 kg when they were not so well fed.

Detailed experiments have been made on this point by Wallace. Ewes after conception were made to grow along the predetermined growth curves shown in Fig. 5.9 by changes in the amount of the rations they

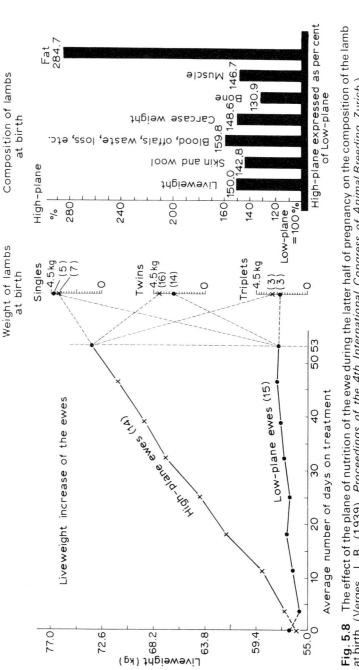

Fig. 5.8 The effect of the plane of nutrition of the ewe during the latter half of pregnancy on the composition of the lamb at birth. (Verges, J. B. (1939). *Proceedings of the 4th International Congress of Animal Breeding, Zurich*.)

received daily. Those killed when 91 days pregnant showed that high or low nutrition in early pregnancy did not affect the weight of the foetus (Fig. 5.10) or development of the udder (Fig. 5.11). After 91 days some ewes were continued on the high and low planes of nutrition, while some in each group were changed over to the other plane (Fig. 5.9). The results in Figs 5.10 and 5.11 show that the plane of nutrition of the pregnant ewe during the last 8 weeks of pregnancy has a great effect on the size of the lamb and development of the udder. High-plane nutrition during this time gives large twin lambs (5.5 kg each) whereas low-plane nutrition gives small lambs (3.4 kg each). Similarly high-plane nutrition gives large udders full of milk and low-plane nutrition udders with little milk. Part of the difference in udder sizes in Fig. 5.11 is due to difference in amount of udder tissue, part to the amount of milk in the gland: in the 'steamed up' ewe,

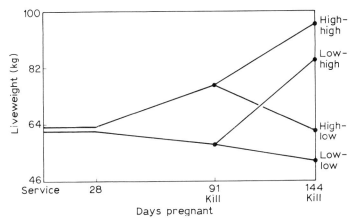

Fig. 5.9 Plan of experiment to find the effect of the plane of nutrition on the pregnant ewe. Liveweight growth curves controlled by individual feeding of the ewes. For results see Figs 5.10 and 5.11. (Wallace, L. R. (1948). *Journal of Agricultural Science*, **38**, 367.)

milk secretion begins earlier. That the milk yield after lambing is greater in ewes well fed in the latter part of pregnancy was shown by Wallace in other experiments. Ewes well fed in late pregnancy averaged a maximum of 23 kg a week, compared with 14 kg from those on a low-plane ration at this time.

As is obvious, the volume of the abdomen increases during pregnancy; but not sufficiently to match the increase in volume of the uterus, particularly in ewes carrying twin lambs. The adjustment between the two is by adjustment in the volume of rumen contents (Forbes, 1969), and consequently by a decrease in the capacity of the ewe to digest fibre. Abdominal fat deposited earlier in pregnancy may also serve to limit rumen capacity. To ensure good lamb size at birth and good growth during suckling, poorer quality fodder should be fed early in pregnancy so that the best, more easily digested and more concentrated, is reserved for the later stages when it can be more profitably fed.

Wallace's experiments show that, particularly when carrying twins, the

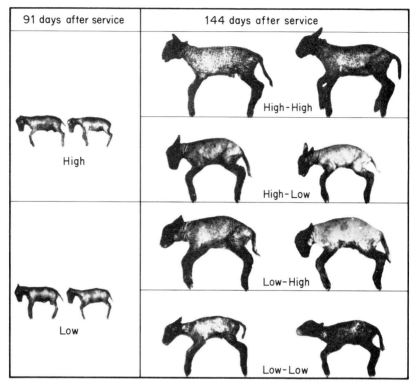

Fig. 5.10   Effect of the plane of nutrition of the ewe during pregnancy (see Fig. 5.9) on the size of the twin lambs (all to the same scale). The plane of nutrition has no effect up to 91 days (*left*) but after that has considerable effect. (Wallace, L. R. (1948). *Journal of Agricultural Science*, **38**, 367.)

ewe cannot supply the needs for growth of the udder tissue and of the lambs from her own muscle and fat, but is dependent on incoming nutrients. As other experiments on sheep (Pálsson; Pálsson and Vérges) and on pigs (Chapter 6) show, each part and tissue of the body competes for the available nutrients, and the tissues have different priorities (see arrows in Fig. 5.12). In early pregnancy the foetal needs are fully met (as shown by the four arrows) but towards the end of pregnancy the foetus comes into competition with maternal tissues. Undernutrition in late pregnancy, particularly in ewes carrying twins, can lead to development of a pregnancy toxaemia, 'twin-lamb disease', in the ewe. It seems that the foetuses compete too effectively with the dam for the carbohydrate available; and that consequently the ewe becomes comatose from the accumulated ketone bodies derived from mobilized body fat (p. 22).

In very hot climates there may be early embryonic loss of lambs, and lambs surviving to birth are relatively dwarfed. The heat-stressed ewes have a low food intake: but, as Yeates has shown, if ewes are fed on a restricted diet so that their weight changes in pregnancy parallel those of ewes under heat stress, their lamb birth weights are not so low as those of

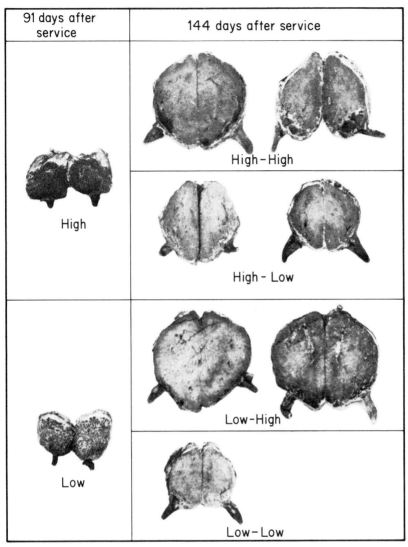

91 days after service	144 days after service
High	High – High
	High – Low
Low	Low – High
	Low – Low

Fig. 5.11   Effect of the plane of nutrition of the ewe during pregnancy (see Fig. 5.9) on the growth of the udder (all to same scale). The plane of nutrition has no effect up to 91 days (*left*) but after that has considerable effect. (Wallace, L. R. (1948). *Journal of Agricultural Science*, **38**, 367.)

the ewes kept in a hot climate. The outcome of the competition between the tissues of ewe and lamb thus are different in different circumstances. The Merino is much less readily affected than most breeds, including some of the hairy desert ones (Yeates, Edey and Hill, 1975).

As in the case of Shire–Shetland crosses (Chapter 3) the size of the placenta is possibly a regulating factor. Egg transplantation has been used to find out how far the size and growth of the lamb is affected by maternal

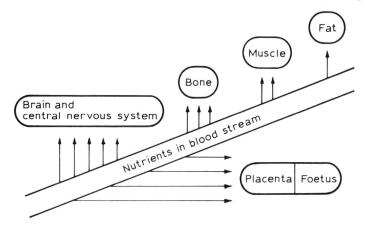

Fig. 5.12 Diagram showing how the different tissues of the body compete for nutrients in the blood stream. Priority of supply varies in the different tissues (denoted by arrows) according to their order of development. When the plane of nutrition is reduced (1 arrow taken from each) fat growth ceases but brain, bone and muscle continue to grow at a slower rate. When the plane of nutrition is reduced still further (2 arrows removed) muscle ceases to grow but fat is now removed to the bloodstream (arrow reversed) to assist in the growth of brain and bone, which still continue to grow but at a slower rate. During early pregnancy the placenta and foetus have a high priority as shown here, but during the last 6 weeks of pregnancy this priority is lost and the foetus is in competition with muscle and fat of the maternal tissue.

Fig. 5.13 A small Welsh ewe and her Welsh lamb (*right*), with a Welsh lamb from an egg transplanted to a larger Border Leicester ewe (*left*); both lambs of the same age (5 weeks). At birth the Welsh lamb born of the large Border Leicester ewe was 0.9 kg heavier than that born of the small Welsh ewe and the difference was increased to 2.7 kg at 5 weeks old.

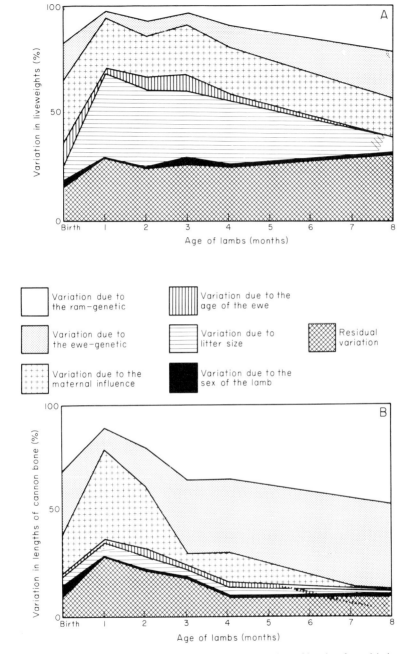

Fig. 5.14 Percentages of the total variation in the size of lambs, from birth to 8 months, which was due to genetics, maternal influence, age of the ewe, litter size and sex of the lamb. **(A)** Liveweight; **(B)** length of cannon bone. (Hunter, G. L. (1956). *Journal of Agricultural Science*, **48**, 36.)

influence as distinct from the genetic influence of the sire and dam (Fig. 5.13). Transfers of fertilized eggs between large Border Leicester and small Welsh ewes have given lambs 0.9 kg heavier at birth in the larger mother. The weight at birth in the larger mother is limited by the genetic capacity of the lamb while in the smaller it is limited by its nutrition. These maternal effects, reinforced as they are by differences in milk supply, last for some time, at any rate until after the age at which prime lamb is usually sold. Figure 5.14 shows the percentage of the total variation in liveweight and length of the cannon bone attributable to the genetics of the sire and dam and to maternal and other influences.

Milk supply and liveweight growth

No one single factor affects the growth rate of the lamb more than the milk supply of the ewe. While the maximum weekly growth rate of single

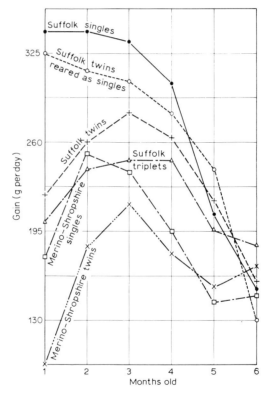

Fig. 5.15 Comparison of the weekly growth rate in single and twin lambs. The maximum growth rate is made in the first month in singles and in twins reared as singles, whereas the maximum growth rate in twins does not occur until the third month owing to limitations in the milk supply. (Hammond, J. (1932). *Growth and the Development of Mutton Qualities in the Sheep*. Oliver & Boyd, Edinburgh.)

lambs is made in the first week of life, the growth rate of twins, owing to limitations of the milk supply, does not reach a maximum until the 5th week when they are beginning to supplement this supply by eating for themselves (Fig. 5.15). This is not due to the difference in size between the singles and twin lambs at birth for, in this respect, twins reared as singles behave as do singles.

In sheep, individuals and breeds vary as much in their milk yield as they do in cattle, and it is important to breed for this quality in ewes which are to be used for producing prime lambs (see Table 11.1). The level of milk supply in a flock can be estimated by observing the difference in size between single and twin lambs: where this ratio is high the ewes are poor milkers (Fig. 5.16).

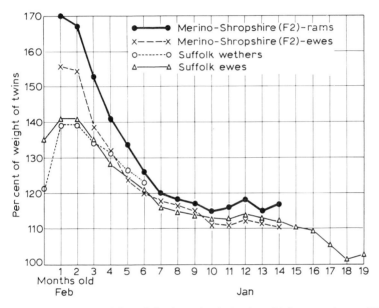

Fig. 5.16 Relative weights of single and twin lambs with increase in age. The ratio between the weights of singles and twins is highest during the suckling period and drops as they begin to eat for themselves. The ratio forms a good index of the milking qualities of the ewe; thus in the Suffolk, a good milking breed, the ratio is lower than in the Merino-Shropshire cross. (Hammond, J. (1932). *Growth and the Development of Mutton Qualities in the Sheep.* Oliver & Boyd, Edinburgh.)

The rate of growth of the lamb is considerably affected by the amount and coarseness of fibre in the food. When a crop like vetches is fed in the young succulent high-protein and low-fibre stage, just before flowering, the lambs will gain weight at the rate of over 2 kg a week; but after the seed-pod stage has been reached, and the plant contains a low proportion of protein and a high percentage of fibre, the rate of gain will drop to 0.5–1.0 kg per week. The addition of concentrates to such a coarse fibrous ration will not raise growth rate to that obtainable on succulent feed, because the more slowly digested coarse feed fills the stomach and restricts appetite.

Market requirements

The most valuable parts of the carcase are the loin and the legs. The loin should be wide, with a well-rounded eye muscle (longissimus dorsi), covered by a thin layer of fat (see Figs. 5.18 and 12.7). The legs should be short-boned and well-filled (U-shaped rather than V-shaped—see Fig. 5.17). The overall compositions of the two carcases illustrated differ very little, (about 15 per cent bone, 55 per cent muscle and 30 per cent fat); but

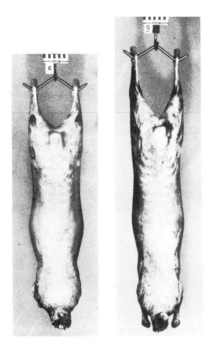

Fig. 5.17 Two New Zealand Export Grade lamb carcases of 14 kg showing the difference in conformation of the Prime (*left*) and Omega (*right*) grades. (Kirton, A. H. and Colmer-Rocher, F. (1978). *World Review of Animal Production*, **14**, 33.)

the conformation of the carcase and the distribution of the fat affect their value. The fat cover, while thin, should spread down the leg—to prevent the meat from drying out in cold storage and in roasting. The deep muscles of the blocky carcase dry out less in cooking, and with the short rib less bone has to be trimmed from the chop before sale.

Lamb fat is normally very firm at room temperature, and this the purchaser expects. Whereas the fat of cattle fattened on cereal is unusually firm (p. 105), sheep fat on such diet tends to be soft and contains branched-chain fatty acids derived from the ruminal fermentation products of starch.

As with other meats, there has been a tendency towards younger carcases which provide smaller joints and which contain less fat. However, as age, size and fatness are linked, it is difficult to know which is really the most important consideration. By skill in cutting, the butcher can to some extent adjust the size of beef joints to what is wanted by the customer. With a leg or shoulder of lamb there is little he can do, except to cut them in halves and, with the development of self-service counters in multiple stores, this is being done increasingly.

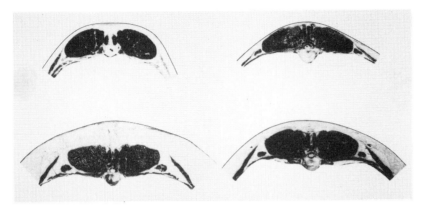

Fig. 5.18 Sections through carcase at the level of the last rib. (*Left*) South-down; at 15 kg (*above*) and at 31 kg (*below*). (*Right*) Blackfaced; at 15 kg (*above*) and 28 kg (*below*). The early-maturing Southdown is more valuable at the lower weight, at which the Blackfaced has more bone and lacks finish. At the heavier weight the Southdown is much too fat; the Blackfaced is also rather too fat, but the eye muscle has now developed and rises above the spine of the backbone. (Hammond, J. (1955). *Journal of the Institute of Meat, London*, No. 11.)

Light lamb carcases are more valuable, and the price per kilo falls as carcase weight increases: but the rate of fall is greater for some breeds than for others because of the difference between breeds in the rate at which they mature—in the sense of reaching desirable proportions of fat, muscle and bone (Fig. 5.18).

The standards for assessment of carcases sent to Britain from New Zealand changed little between 1952 and 1966 (Table 5.3) with the change in demand, in spite of the desire for less fat, and of the tendency to self-service sales of small joints—which makes overall carcase appearance, described above, of less importance than formerly.

Development of body proportions

The wild sheep, the Mouflon, changes in the proportions of its body as it grows up, but these changes do not proceed so far or as fast as in those sheep selected for meat production (Fig. 5.19). At birth the lamb is all head and legs, but as it grows up it lengthens and deepens, so that the proportion of more valuable parts such as the loin and rump are increased relative to

Table 5.3 Percentage of bone, muscle and fat in New Zealand export lamb 1952 and 1966. (From Clarke, F. A. and McMeekan, C. P. (1952). *New Zealand Journal of Science and Technology*, **33**, 1; Kemp, J. D. and Barton, R. A. (1966). *New Zealand Journal of Agricultural Research*, **9**, 590.)

Grade	1952				1966			
	Average weight (kg)	Bone & tendon %	Muscle %	Fat %	Average weight (kg)	Bone & tendon %	Muscle %	Fat %
Down 2	14.5	14.4	50.5	33.6	15	12.9	54.0	32.7
Prime 2	14.5	16.3	56.0	26.5	15	14.0	53.3	32.2
3	18	15.7	53.2	30.2	–	–	–	–
4	21	14.7	52.8	31.5	–	–	–	–
YM	14	18.9	56.8	22.1	15	15.5	56.9	27.0

the less valuable parts such as the neck and shanks. As with cattle, there is an optimum point beyond which deepening of the body results in excessive waste in the carcase, due to deep ribs and flank, and where there is excessive fattening. The body proportions of the adult Mouflon ewe develop to only about the same extent as is reached by the improved Suffolk breed at about 3 months old.

In the lamb at birth, little more than half the liveweight is carcase; and when the bone is removed only 31 per cent remains as edible meat. At 22 months old the proportion which is edible is twice as great (Table 5.4). However, the composition of what is 'edible' does not remain unchanged; the proportion of fat increases, and the proportion of muscle in the carcase actually decreases after a while. Development in improved breeds proceeds so far, under good nutritive conditions, that in the adult there is far too much fat. Desirable body proportions are reached while growth is still rapid (and hence while food conversion rate is efficient). This point, in the case of the Suffolk, is at an age of about 4 months, with a carcase containing 20 to 25 per cent fat.

These changes in body proportions and composition are brought about by differential growth gradients existing between the different parts and tissues of the body. Thus in the leg after birth (Fig. 5.20) a gradient from low to high rate of growth runs from the lower part such as the cannon

Table 5.4 Changes in conformation of the sheep as it grows up. (From Hammond, J. (1932). *Growth and the Development of Mutton Qualities in the Sheep*. Oliver & Boyd, Edinburgh.)

Suffolk sheep	Birth	3 months	11 months	22 months
100 kg of live weight contain { kg carcase	53	54	60	67
kg edible muscle and fat	31	42	54	62
kg edible muscle	30	36	34	31
kg edible fat	1	6	20	31
kg bone	17	9	5	4
% fat in leg of mutton	2	5	20	30

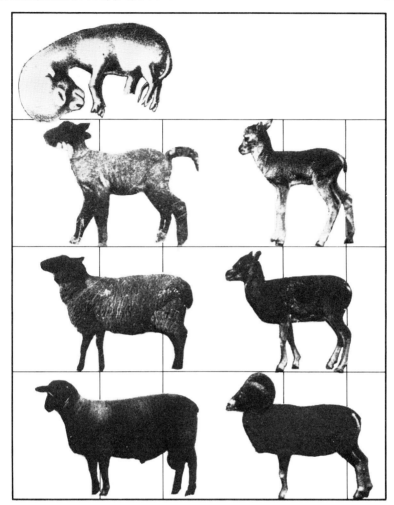

Fig. 5.19 Changes in the proportions of the improved Suffolk (*left*) and unimproved Mouflon (*right*) breeds of sheep as they grow up. Each is reduced to the same shoulder height so as to show differences in proportions as distinct from differences in actual size. *Top*, embryo of 2 months; *2nd line*, 4 days old; *3rd line*, adult ewe; *bottom*, adult ram. (Hammond, J. (1932). *Growth and the Development of Mutton Qualities in the Sheep*. Oliver & Boyd, Edinburgh.)

bone to the upper part such as the femur; so that, while at birth in the Suffolk the femur is only 217 per cent of the cannon bone weight, in the adult it is 324 per cent. Similarly there is a gradient between bone, muscle and fat in the leg, so that while at birth the muscle is only about 159 per cent and the fat 9 per cent of the bone weight, in the adult the muscle is 560 per cent and the fat goes up to 285 per cent. A quick rate of change in these proportions is what is meant by early maturity from the butcher's point of view. There are considerable breed differences in the rate at which these

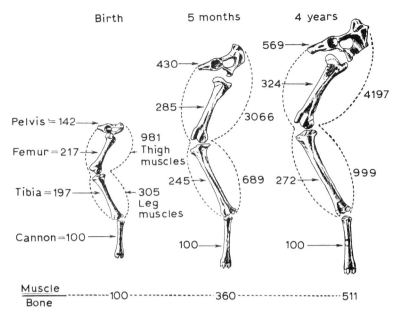

Fig. 5.20 Age changes in the proportions of the parts in the leg of mutton of Suffolk ewes. In each case the weights of the bones and muscles are shown as a percentage of the weight of the cannon bone. (Hammond J. (1932). *Growth and the Development of Mutton Qualities in the Sheep.* Oliver & Boyd, Edinburgh.)

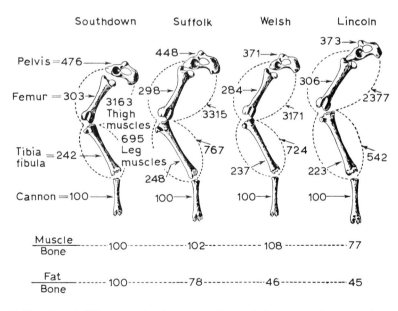

Fig. 5.21 Breed differences in the proportions of the parts of the leg in wethers at 5 months old. In each case the weights of the bones and muscles are shown as a percentage of the weight of the cannon bone. (Hammond, J. (1932). *Growth and the Development of Mutton Qualities in the Sheep.* Oliver & Boyd, Edinburgh.)

changes occur (Fig. 5.21). It is easier to obtain early maturity in small breeds than in large ones under similar feeding conditions; thus, at 5 months old, in the leg of the small early-maturing Southdown the muscle is about 485 per cent and the fat 266 per cent of the bone weight; in the large late-maturing Lincoln the muscle is only about 366 per cent and the fat 99 per cent of the bone weight.

For prime lamb production the breed and type of ram used is determined by the market requirement on one hand, and by the availability of feed on the other. If the season of good green feed is short—3 or 4 months—as in many areas of Australia which are marginal for lamb production, a short-legged early-maturing breed such as the Southdown is used. Cross-bred lambs sired by such rams reach the required body proportions of bone, muscle and fat within the time limit set by the annual pasture, at very light carcase weights—13 to 15 kg. Even at weights as light as 17 to 18 kg such carcases generally will be overfat for the modern market (Table 5.3). Hence, if the season is longer, a somewhat later-maturing breed such as the Dorset Horn is used. The time to reach the required body composition will be a little longer and the carcases will be heavier—16 to 18 kg—but not overfat. Where lambs of even higher weight are required and where the grass season is sufficiently long—5 to 6 months—it is better to use a larger ram such as the Suffolk.

It is generally most efficient to grow lambs on a high plane of nutrition from birth to slaughter and to select breeds suitable for the environment, and for the market at which one is aiming. If growth is slowed, optimum proportions are attained at a higher carcase weight (see below).

Plane of nutrition and development

Each part and tissue of the body follows the same basic pattern of growth. The rate at which each grows at first increases and then slows; but they reach their peak rate of growth at different times, and those which are earliest developing form a greater proportion of the whole earlier in life. Bone is earlier developing than muscle, so muscle is proportionately greater later in life (Fig. 5.20). Fat is the latest developing tissue, and also the one most severely affected by reduced food intake (and hence reduced growth rate). On a low plane of nutrition maturity is delayed.

In commercial carcases (Table 5.5), the average chemical composition on a fat-free basis is remarkably similar over a wide range of ages and weights. The lower percentage of water and higher percentage of protein in the frozen carcases is due to evaporation loss. The percentage of fat is largely determined by age and plane of nutrition.

Figures 5.22 and 5.23 illustrate an experiment on the effect of carcase composition of extreme levels of feeding. Figure 5.22 shows the plan of the experiment, and Fig. 5.23 the respective carcase proportions at the same carcase weight, but on different patterns of growth.

On a high plane throughout (HH) the lamb is more mature, in terms of fatness, at an early age and at low carcase weight. The LH animal, which had bone and muscle growth slowed during early life and was then finished

Table 5.5   Carcase and body composition of New Zealand ewes and lambs. (Data from Kemp, J. D. and Barton, R. A. (1966). *New Zealand Journal of Agricultural Research*, **9**, 590; Kirton, A. H., Ulyatt, M. J. and Barton, R. A. (1959). *Nature, London*, **184**, 1724.)

Type of carcase Group	Age (months)	Average carcase weight (kg)	Percent fat	Percent composition on fat free basis		
				Water	Protein	Ash
A. Frozen carcases						
Down D	under 8	12	31	72	22	6
2	under 8	15	33	73	21	6
Prime D	under 8	12	28	72	22	6
2	under 8	15	32	73	21	6
YL	under 8	11	25	71	23	6
YM	under 8	14.5	27	73	22	5
Alpha	under 8	9.5	20	72	22	6
B. Hot carcases						
Crossbred lambs						
(heavy)	7	18	32	74	20	6
Ewes (average)	27	29	42	74	20	6
Ewes (fat)	over 60	51	56	73	21	6
Ewes (thin)	over 60	14	17	75	19	6

rapidly, attained a comparable degree of maturity at the same weight but at twice the age. Of the remaining two, the HL animal made more bone growth early: it reached the same weight at the same age as the LH lamb but was longer, deeper and leaner, i.e. less mature. The LL animal ultimately attained a similar weight and carcass conformation but the age was doubled again.

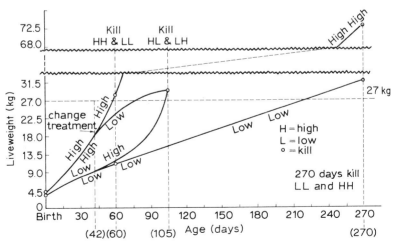

Fig. 5.22   Growth curves of lambs controlled by feeding on different planes of nutrition. The effect of this on the proportions of the carcase is shown in Fig. 5.23. (Verges, J. B. (1939). *Suffolk Sheep Society Yearbook*. Ipswich.)

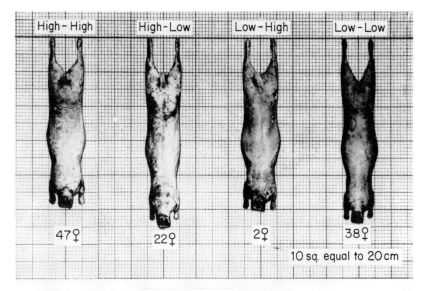

Cut between last two ribs

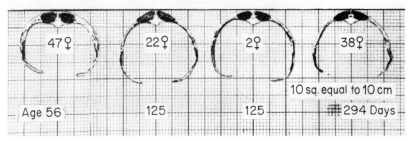

Fig. 5.23 The effect of the shape of the growth curve (see Fig. 5.22) on the carcase proportions of lambs at the same carcase weight (13.5 kg). (Verges, J. B. (1939). *Suffolk Sheep Society Yearbook*. Ipswich.)

The skeleton as an index of type

Since the bones have a high impetus for growth early in life they suffer less than do muscle and fat under conditions of poor nutrition and so can be used as indices of inherent type.

The bones of the improved meat breeds are shorter and relatively thicker than those of unimproved types, or of the improved wool breed, the Merino (Fig. 5.24). Increased thickness of bone is accompanied by increased depth of flesh. Similarly the head in the improved meat-types is wide with a short face. Thickness of bone and increased muscularity is a characteristic of the male. As bones grow they change in shape, just as does the whole animal. The bone grows in width after it ceases to grow in length, and the shape changes (Fig. 5.25). In the male there is greater thickness growth of the bone than in the female or castrate (Fig. 5.26).

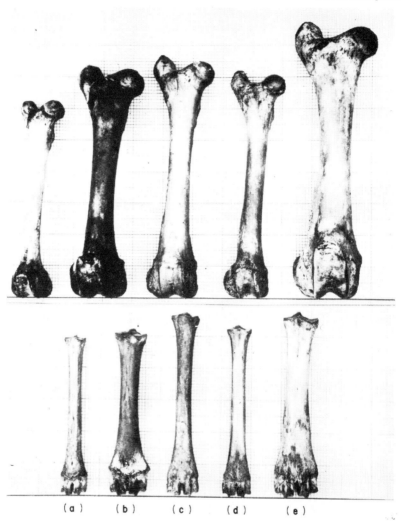

(a)   (b)   (c)   (d)   (e)

Fig. 5.24 Differences in the shape of the hind cannon bone and femur in rams of different breeds of sheep. Breeds unimproved for mutton have slender bones similar to those of the young of improved breeds (see Fig. 5.25). **(a)** Soay. **(b)** Hampshire. **(c)** Merino. **(d)** Shetland. **(e)** Suffolk. (Hammond, J. (1932). *Growth and the Development of Mutton Qualities in the Sheep.* Oliver & Boyd, Edinburgh.)

When lambs are reared on a low plane of nutrition the bones are slender compared with those reared on a high plane (Fig. 5.27) just as are those of unimproved as compared with improved breeds.

The spinous processes of the vertebrae are short in meat-type animals, so in the well-grown animal the 'eye' or longissimus dorsi muscle of the chop stands out above the bone (Fig. 5.28). The length of the spinous

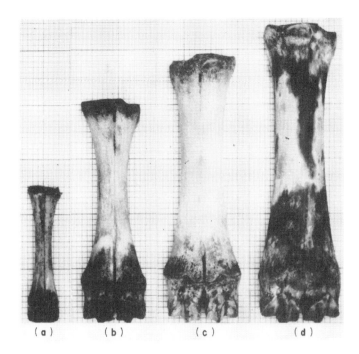

Fig. 5.25 Age changes in the shape of the fore cannon bone in Suffolk rams; all to the same scale. There is a gradual increase in relative thickness. **(a)** 2 days; **(b)** 3 months; **(c)** 5 months; **(d)** 4 years. (Hammond, J. (1932). *Growth and the Development of Mutton Qualities in the Sheep.* Oliver & Boyd, Edinburgh.)

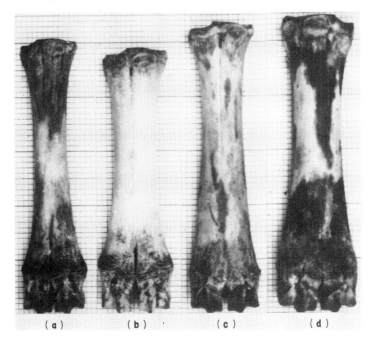

Fig. 5.26 Sex differences in the thickness of the fore cannon bone in Suffolk sheep; all to the same scale. **(a)** Ewe and **(b)** ram at 5 months old; **(c)** ewe and **(d)** ram at 4 years old. (Hammond, J. (1932). *Growth and the Development of Mutton Qualities in the Sheep.* Oliver & Boyd, Edinburgh.)

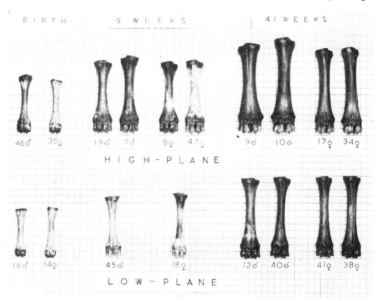

Fig. 5.27 The effect of the plane of nutrition on the growth of bone in lambs at different ages. A low plane of nutrition affects the late-developing thickness growth of bone much more than it does the early-developing length growth, and the larger wether ($\male$) than it does the smaller ewe ($\female$). (Palsson, H. and Verges, J. B. (1952). *Journal of Agricultural Science*, **42**, 1.)

processes is most pronounced over the shoulder, for here the ligamentum nuchae which supports the head is attached. When the neck is shortened and the head is light the spinous processes are shortened and a broad rather than a sharp shoulder is formed. Thus a small head and neck and broad shoulders go together with short cannons and 'well led down' legs, and are used traditionally in judging the value of an animal. However, the shoulders of Cheviots are higher and more pointed than those of Romneys. This is due to the slope of the spinal column, to the height of the spine, and to less fat and more muscle in the shoulder region of the Cheviot. Excessive emphasis on short bones, and width and roundness over the shoulder, has led to the development of sheep with excessive fat in the forequarter.

Development of hair and wool

Hair is a characteristic feature of all mammals: wool is a modified type of hair. The hair root is a down growth from the skin and the hair grows up, through a sheath continuous with the root, into a pit on the skin surface. An ordinary hair root is not normally active continuously; the hair grows for a while, then the hair follicle goes into an inactive phase. After a period the follicle becomes active again, a new hair grows and the old one is shed. The human scalp follicle, and most wool follicles, are peculiar in

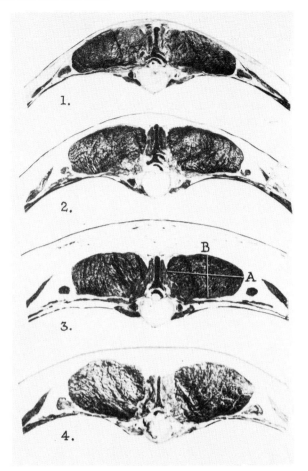

Fig. 5.28   Cross-sections of the longissimus dorsi muscle of wethers, cut at the level of the last rib, to show differences in the shape index $\dfrac{B \times 100}{A}$ of the muscle and the length of the spinous process in relation to this.

	A (mm)	B (mm)	Shape index
1. Blackfaced	52	24	46
2. Suffolk	65	35	54
3. Hampshire	58	33	57
4. Southdown	62	43	69

(Hammond, J. (1936). *Festschrift Prof Duerst*, Bern.)

having very long periods of growth (several years) and very brief resting periods. In many species the activities of the hair follicles are hormonally coordinated so that there are seasonal moults.

Hairs are generally pigmented, and in many animals there are two types of hair in the coat. The outer coat is of long coarse guard hairs (primaries) while below there are more numerous short fine fur hairs, or secondaries, which retain an insulating blanket of still air next to the skin surface. Fur hairs are solid cylinders of keratin; guard hairs—except at their tips and bases—are generally medullated. That is, they consist of tubes of keratin with a central airspace broken up by thin partitions. Medullated fibres are not optically homogeneous (air and keratin having different refractive indices) and are commonly described as 'chalky' in appearance. The hair is not smooth, but has a scaly surface, which is well developed in wool and confers the capacity to felt. Wool fibres differ from most hairs in that they do not grow straight, but with a characteristic waviness (crimp) the frequency of which is characteristic of the breed (see Fig. 5.32). This crimping in some way is dependent on copper, and loss of crimp in wool is a sensitive indicator of dietary deficiency (or unavailability) of copper.

The wool sheep has been evolved from an animal with a pigmented double coat which had seasonal moults. The primary follicles are characteristically associated in groups of three, and can be distinguished from secondary follicles because they, but not the secondaries, each have a sweat gland associated with them. Improvement has consisted, in different degrees in different breeds, in the elimination of pigment, increasing the density of the skin follicle population, decreasing relative length and diameter of primary fibres (with elimination of medullation), increasing the ratio of secondaries to primaries (S/P ratio) and elimination of moulting. Some breeds have complete seasonal moults (e.g. the Wiltshire Horn) and in others some primaries may go into rest and are shed subsequently, these fibres being shorter than those of the rest of the fleece.

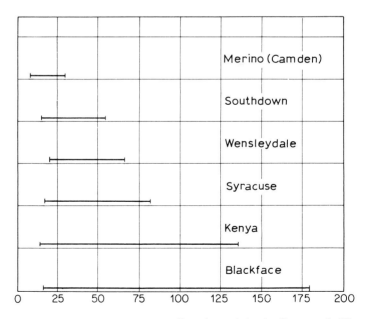

Fig. 5.29 Range of fibre diameters (in microns) in the fleeces of different types of sheep. (From Barker, A. F. (1925). *Report of the 24th International Conference; National Sheep Breeders' Association, London.*)

The fleeces of British breeds may be classified into three main types, carpet wool, down type and long wools. Carpet wools come from double-coated mountain breeds such as the Blackface and have a great range of fibre diameters (Fig. 5.29). A long coarse outer coat seems to be an adaptation to survival in a cold damp climate. In the long wools (e.g. the Lincoln and Romney) the coat is long and not very dense; in the down or short-wool type (e.g. Southdown) it is dense but shorter; in both there is no absolute size distinction between fibres from primary and from secondary follicles. The finest wools are produced by Merinos, which have a very dense fleece with a high S/P ratio, and in which the primary fibres have been much reduced in diameter (Fig. 5.29).

In the foetus the primary follicles develop first, and generally only primaries develop on the face and legs. Development of primaries is nearly completed before secondaries appear. The time at which secondaries first appear, and the period of time during which secondary follicles can continue to be formed, differs between breeds, and the extent of their development can be limited by the environment. Thus Weiner and Slee's transplantation of eggs between Welsh Mountain and Lincoln ewes has demonstrated a maternal influence on the birthcoat, and Schinckel showed that lambs born as twins, and from poorly fed ewes, have lowered adult S/P ratios. Good early nutrition is necessary to develop full genetic potential for density of fleece.

The birth coat is shed within a few months of birth, and is not fully representative of the adult fleece—thus, for example, the Suffolk lamb may be pigmented, and the Romney may have long coarse (primary) halo hairs (see Fig. 11.2) which in the adult fleece are replaced by finer fibres. After all the secondary follicles are formed the fleece density decreases because of expansion of skin surface area as the lamb grows (Fig. 5.30); in general, the higher the S/P ratio the finer and denser is the fleece (Figs. 5.31, 5.32).

There are skin gradients in fleece development. Generally S/P ratio is greatest and fleece finest over the shoulder, and the coarsest fibres are found on the lowest part of the leg, the britch. Uniformity of the fleece is an advantage. Halo hairs in the Merino birthcoat and hairiness of the Romney lamb's tail have been shown to predict poor uniformity of adult fleece.

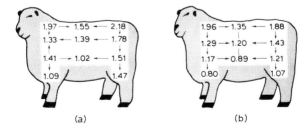

(a)   (b)

Fig. 5.30 Percentage of skin area occupied by wool fibres in Romney lambs during the period 4 to 20 weeks (*left*) and 20 to 36 weeks (*right*). The reduction in density is a measure of skin growth in the different areas. (Henderson, A. E. (1953). *Journal of Agricultural Science*, **43**, 12.)

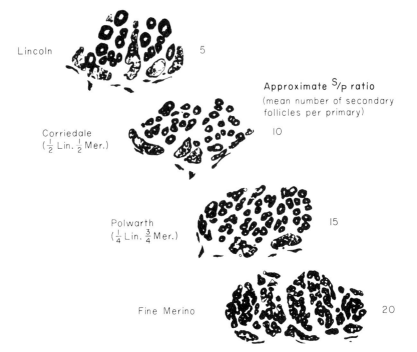

Lincoln

Corriedale
($\frac{1}{2}$ Lin. $\frac{1}{2}$ Mer.)

Polwarth
($\frac{1}{4}$ Lin. $\frac{3}{4}$ Mer.)

Fine Merino

5

Approximate $^S/_P$ ratio
(mean number of secondary
follicles per primary)

10

15

20

Fig. 5.31 Wool follicles in sections of the skins of different breeds and crosses of sheep. Each group contains 3 primary follicles but very different numbers of secondary follicles. There are differences in the length, density, fineness and number of crimps per cm as shown in Fig. 5.32. (Carter, H. B. (1957). *Australian Journal of Agricultural Research*, **8,** 109.)

Mean fibre diameter (approx)

46μm   33μm   27μm   20μm

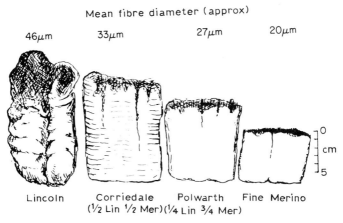

Lincoln   Corriedale   Polwarth   Fine Merino
(½ Lin ½ Mer)(¼ Lin ¾ Mer)

0
cm
5

Fig. 5.32 Ten months' growth of wool under the same high-plane nutritional conditions from the different breeds and crosses of which the wool follicles are shown in Fig. 5.31. In the Lincoln the wool is long, not dense, coarse in fibre, and with only about 1.2 crimps per cm; in the Merino it is short, dense, fine and with about 9.5 crimps per cm; crosses are intermediate. (Carter, H. B. (1957). *Australian Journal of Agricultural Research*, **8,** 109.)

Growth in length of the fibres, and also their diameter, is reduced under poor nutritive conditions (Marston, 1955). The finest wool is produced from Merinos kept under marginal conditions. The wool sheep is unique in its capacity to yield a useful product under submaintenance conditions and the Merino is kept in many dry areas of the world where meat production is impracticable.

Although seasonal variation in nutrition is the major factor affecting the annual rhythm of wool growth, there is also an effect of photoperiod (Coop and Hart, 1953). Rate of growth is greatest in the summer months. The effect is less marked in the Merino than in breeds originating further from the tropics. Stress, such as severe infection, can produce localized thinning of the wool follicle. Unlike the sustained thinning due to poor nutritive conditions, this is a serious defect, because the fibres are liable to break at the weak point during manufacture.

References

BUTTLE, H. L., COWIE, A. T., JONES, E. A. and TURVEY, A. (1979). Mammary growth during pregnancy in hypophysectomised or bromocriptine-treated goats. *Journal of Endocrinology*, **80**, 343.

CORTEEL, J. M. (1974). Viabilité des spermatozoïdes de bouc conservés et congelés avec ou sans leur plasma seminal: effet du glucose. *Annales de Biologie Animale Biochemie Biophysique*, **14**, 741.

CORTEEL, J. M. (1975). The use of progestagens to control the oestrous cycles of the dairy goat. *Annales de Biologie Animale Biochemie Biophysique*, **15**, 353.

COLAS, G. (1975). Effect of initial freezing temperature, addition of glycerol and dilution on the survival and fertilizing ability of deep frozen ram semen. *Journal of Reproduction and Fertility*, **42**, 277.

COOP, I. E. and HART, D. S. (1953). Environmental factors affecting wool growth. *Proceedings of the New Zealand Society of Animal Production*, **13**, 113.

EVANS, G. and ROBINSON, T. J. (1980). The control of fertility in sheep: endocrine and ovarian responses to progestagen—PMSG treatment in the breeding season and in anoestrus. *Journal of Agricultural Science*, **94**, 69.

FORBES, J. M. (1969). The effect of pregnancy and fatness on the volume of rumen contents in the ewe. *Journal of Agricultural Science*, **72**, 119.

KNIGHT, T. W. and LYNCH, P. R. (1980). Source of ram pheromones that stimulate ovulation in the ewe. *Animal Reproduction Science*, **3**, 133.

LINDSAY, D. R. (1966). Mating behaviour of ewes and its effect on mating efficiency. *Animal Behaviour*, **14**, 419.

MARSTON, H. R. (1955). Wool growth. In *Progress in the Physiology of Farm Animals* (J. Hammond, Ed.). Butterworth, London.

SALAMON, S., MAXWELL, W. M. C. and FIRTH, J. (1979). Fertility of ram semen following chilled storage (5°C). *Animal Reproduction Science*, **2**, 373.

TRAPP, M. J. and SLYTER, A. L. (1979). Pregnancy diagnosis in the ewe. *South Dakota State University Agricultural Extension Service*, **79**, 11.

YEATES, N. T. M. (1949). The breeding season of the sheep with particular reference to its modification by artificial means using light. *Journal of Agricultural Science*, **39**, 1.

Further reading

HAMMOND, J. (1932). *Growth and the Development of Mutton Qualities in the Sheep.* Oliver and Boyd. Edinburgh.

HUNTER, R. H. F. (1980). *Physiology and Technology of Reproduction in Female Domestic Animals.* Academic Press, London and New York.

ROBINSON, T. J. (Ed.) (1967). *The Control of the Ovarian Cycle in the Sheep.* Sydney University Press.

SALAMON, S. (1976). *Artificial Insemination of the Sheep.* Department of Animal Husbandry, University of Sydney.

YEATES, N. T. M., EDEY, T. N. and HILL, M. K. (1975). *Animal Science.* Pergamon Press, Sydney.

6 Pigs

The oestrous cycle

The domestic pig exhibits oestrous cycles the whole year round, but some wild pigs living in extreme environments show evidence of seasonal breeding. The normal time between one heat and the next is 21 to 22 days (Fig. 6.1), while the actual heat period lasts from 40 to 65 hours and is rather shorter in gilts than in sows. Heat is marked by a swelling of the vulva; there is a gradual increase in size from about the 9th day before heat, and a decrease for about 9 days after. There is a rapid increase in the level of oestrogen (mainly oestrone) excreted by the sow; this starts on the 15th day of the cycle and reaches a peak just before the onset of heat, after which it falls (Fig. 6.2). The oestrogen is responsible for the swelling of the vulva and the associated changes in the vaginal mucus.

In heat periods of 40 to 60 hours' duration, ovulation appears to occur about 18 hours before the end of the period of oestrus. McKenzie made double matings of a Saddleback sow—with a Saddleback boar at the 12th

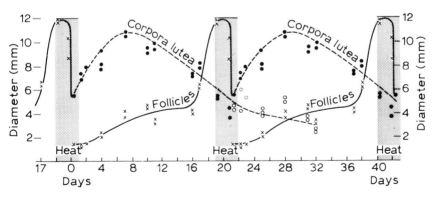

Fig. 6.1 Diagram showing some of the cyclic changes occurring in the ovary of the sow. Changes in the average size of the largest set of follicles are shown; these reach a maximum during heat. Rupture then occurs, the follicles collapse, and the corpora lutea formed from them reach their maximum size seven days after heat, after which they regress. (McKenzie, F. F. (1926). *Research Bulletin of the Missouri Agricultural Experiment Station*, No. 86.)

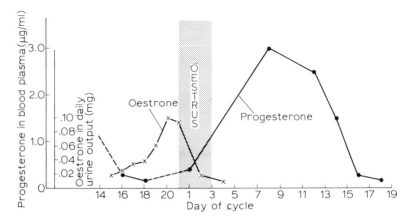

Fig. 6.2 Changes in production of ovarian oestrogen and progesterone during the oestrous cycle of the pig. (Drawn from data of Bowerman, A. M., Anderson, L. L. and Melampy, R. M. (1964). *Iowa State Journal of Science*, **38**, 437; and of Masuda, H., Anderson, L. L., Henricks, D. M. and Melampy, R. M. (1967). *Endocrinology*, **80**, 240.)

hour of heat and with a White boar at the 36th—and found that young were produced from the mating at the 12th hour. Matings on the second day of the heat result in a higher percentage of fertilized eggs than do matings on the first or third day (Fig. 6.3), while matings after the end of heat result in almost complete failure. The corpus luteum reaches its maximum function in 8 days (Fig. 6.2) and, if the sow is not pregnant, regression commences after the 14th day.

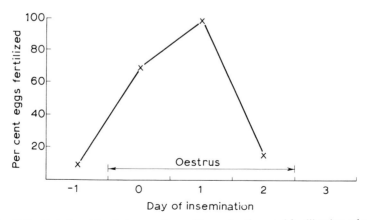

Fig. 6.3 Relationships between time of insemination and fertilization of eggs in the sow. (Drawn from data of Hancock, J. L. and Hovell, G. J. R. (1962). *Animal Production*, **4**, 91.)

Fertility and sterility

Infertility in gilts is most commonly due to anatomical faults in the reproductive tract, and in sows to early death of the embryos (Table 6.1). There
can be both partial and complete loss of litters (in the animals of Table 6.2,
for example, there appears to have been total loss of several litters between
about one and four weeks after insemination). The number and position
of embryos in the uterus affects their chance of survival. If by day 12–14
more than half the available uterine space is unoccupied by developing
embryos, pregnancy terminates (Dhindsa and Dzuik, 1968), and at least
four embryos appear necessary to maintain pregnancy at that stage.

Partial loss (see Fig. 1.4) is common, and the number of eggs shed by
sows does not normally constitute a factor limiting the number of young
born, which appears to be determined rather by a capacity of the sow to

Table 6.1 Causes of reproductive failure in repeat-breeder gilts and sows. (Warnick,
A. C., Grummer, R. H. and Casida, L. E. (1949). *Journal of Animal Science*, **8**, 569.)

Reason for failure	Gilts %		Sows %
Failure of fertilization			
Gross defects in tract	50.0		15.8
No gross defects	3.4		16.8
Total		53.4	32.6
Embryonic death		23.9	67.4
Ultimately conceived		22.7	0.0
Total		100.0	100.0

carry to term only a limited, smaller, number of young. It seems that this
may be inherited in much the same way as in the rabbit (see page 258).
However it has been suggested that immunological incompatibility between sow and piglet may also be involved, and American data (Dzuik,
1977) indicate that the boar used can be an important factor.

Cystic follicles are a common cause of sterility but are less important
than total embryonic loss, and are most common in old sows (Pomeroy,
1960). They do not respond readily to injections of luteinizing hormone.

Although the number of eggs shed by sows does not normally constitute
a factor limiting the number of young born, as is the case in sheep and
cattle, the number in gilts is frequently low. 'Flushing' of gilts—feeding for
2 or 3 weeks on a high-energy supplement—will increase the number of
eggs shed (Zimmerman, Spies, Self and Casida, 1960). However there is
also an increase in the percentage of embryonic mortality. There is evidence
that reduction of the level of feeding after service may reduce embryonic
loss (see Table 6.4). Treatment with MSG while the follicles are growing
will increase the number of ovulations, as in sheep and cattle, but again
there is a great loss of developing embryos in the first 25 days of pregnancy
(Hunter, 1966).

Sows differ from ewes and cows in that puberty—the time when oestrous cycles commence—seems to be more dependent on age than upon weight. Gilts grown on a moderate plane of nutrition will breed at the same age as those well fed. It is necessary to restrict feed intake by 50 per cent to delay the age of mating. Rearing on a moderate plane of nutrition will ensure optimum reproductive performance and will prevent breeding difficulties associated with overfeeding (Lodge and McPherson, 1961).

The boar and artificial insemination

At mating the tip of the penis engages with the cervix so that ejaculation takes place directly into the uterus. Copulation is prolonged, and lasts from 3 to 10 minutes or more. Ejaculation is an intermittent process, and three fractions can be recognized in the flow of semen: a clear pre-sperm portion is followed by a larger volume rich in sperm, and this is succeeded by a gelatinous fraction which amounts to about half of the total ejaculate—the total volume of which amounts to 140-400 ml. Sperm are concentrated within the uterus by absorption of accessory fluids. Over frequent use of the boar will result in reduced fertilization of ova, and consequently small litter-size.

Semen for artificial insemination is collected with an artificial vagina, simply by grasping the penis with a gloved hand and using a sheath or cone (Fig. 6.4) to collect the semen. The boar works well with a dummy female, and an oestrous sow is rarely required. For artificial insemination, the pre-sperm fraction is discarded and the gelatinous material is filtered off.

The sperm are very sensitive to temperature shock, so semen is collected into a warmed container. It can be stored (diluted 1:2 or 1:5) for two or three days at 15-18 °C, and further diluted for insemination to give a dose of 2×10^9 sperm in a volume of 50 ml. The semen is delivered into the uterus either by forcing it through under pressure from the anterior end of the vagina, or by use of a spirally grooved rubber catheter which, like the boar's penis, will engage with the cervix.

Various workers, mostly using the pellet method (p. 75), have obtained pregnancies from frozen semen, with conception rates of from 30 to 75 per cent (Larsson, 1978). However, so far the number of sperm required for successful insemination has been very much greater than with fresh semen; consequently the technique does not have wide commercial usefulness.

Reported farrowing rates from artificial insemination range from 60 to 85 per cent, with results for gilts generally poorer than for sows, but normal litter size if sufficient sperm are used. The main reason for poor results seems to be difficulty in accurate detection of oestrus in the absence of a boar. In France, Signoret (1972) reported a conception rate of only about 58 per cent, but when sows were inseminated in the presence of a boar, and exhibited a full mating reflex, the rate increased to 75-80 per cent.

The reflex (Fig. 6.5) is facilitated by the sight, sound, and particularly by the scent of the boar. The substance responsible is a male hormone derivative, which is now available in aerosol cans so that it can be sprayed

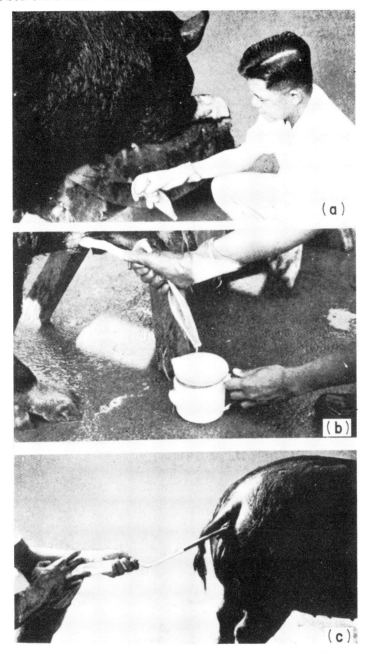

Fig. 6.4 (a) Boar mounting a dummy in preparation for semen collection. (b) Semen collection from boar by use of a smooth rubber liner and continuous pressure exerted by hand near distal end of penis. (c) Artificial insemination of sow. (Trimberger, G. W. (1962). In *Reproduction in Farm Animals*, (E. S. E. Hafez, Ed.) Lea & Febiger, Philadelphia.)

Fig. 6.5 Mating reflex in the sow. (*Above*) Sow in heat standing rigid when pressure is applied to the region of the loin. (Photograph by courtesy of J. P. Signoret.) (*Below*) Sow giving a positive response to the riding test for oestrus. (Melrose, D. R. (1966). *World Review of Animal Production, special issue*, **2**, 15.)

around the room in which the sows are to be tested. Tape recordings of the boar mating cries are also sometimes used.

Artificial control of breeding

Surgical recovery (Fig. 6.6) and transfer of fertilized ova is practicable; under optimal conditions there is 60-70 per cent survival of transferred embryos. It is also possible to induce heat, followed by ovulation, during lactation; Heitman and Cole (1956) found that if injections of MSG were given between the 20th and 39th day of lactation, 76 per cent of the sows came on heat and, of these, 44 per cent (one-third of the total treated) produced litters. Later in lactation, between the 40th and 50th days, 86 per cent came on heat and, of these, 66 per cent (i.e. about a half of the total treated) produced litters.

However the main practical benefit to be obtained is from synchronization of oestrus, for the purposes both of subsequent artificial insemination and of ensuring that farrowing occurs in batches, over short periods of time.

In principle, the methods applicable with sheep and cattle can be used with pigs also. In practice, the use of prostaglandins to cause regression of

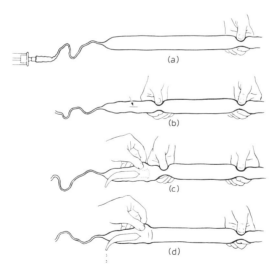

Fig. 6.6 Stages in the surgical recovery of eggs from the uterus of the sow (not to scale). **(a)** The fluid is injected via the Fallopian tube. **(b)** The fluid is held by Operator 2 within a length of uterus clamped with the fingers at each end; the uterus is incised at a point, arrowed, proximal to the distended length. **(c)** A cannula is inserted through the incision and is held in position by Operator 1. **(d)** Operator 2 releases the clamp near the cannula to allow the fluid contents to escape from the cannula. (Hancock, J. L. and Hovell, G. J. R. (1962). *Journal of Reproduction and Fertility*, **4**, 195.)

the corpus luteum is limited by the fact that the corpus luteum is not sensitive until the latter part of the cycle. This can be overcome if a group of gilts are allowed to become pregnant from service at natural heat, and are then treated not less than 12 days after the last one was served (Table 6.2). Heat generally occurs 4–7 days later. As with sheep just before the start of the breeding season, introduction of a male to a group of females tends to advance the time of first ovulation, and to some extent to synchronize the timing of ovulations.

Synchronization by treatment with progesterone to prolong the cycle (suppressing ovulation until after the end of the period of treatment) has been attempted, but subsequent fertility has been poor with, quite commonly, development of cystic follicles; very possibly this was due to the dose level of progesterone having been too low.

Feeding a synthetic substance, methallibure, will suppress pituitary gonadotrophic activity in pigs. On cessation of a 20 day period of feeding the

Table 6.2 Fertility of gilts inseminated at heat after treatment with prostaglandin (and consequent termination of any existing pregnancy). (Guthrie, H. D. and Polge, C. (1978). *Journal of Reproduction and Fertility*, **52**, 271.)

Time of embryo recovery (days)	Number of gilts			Mean no. of	
	Treated	Inseminated	With embryos at slaughter*	Corpora lutea	Embryos
4–7	30	28	26 (93%)	13.5	12.6
25–32	30	24	18 (75%)	14.8	12.2
Total	60	52	44 (85%)	14.0	12.4

*Percentages are of gilts inseminated

majority of gilts come into heat in 5 to 7 days, and matings are fertile. The timing of ovulation can be made more precise by treatment, in addition, with MSG and HCG; the technique is to feed the compound for 20 days, inject MSG on day 21 and HCG 4 days later, and to inseminate on the day following. It has not been found practicable to control ovulation in sheep and cattle by feeding methallibure. The substance has been found capable of causing impaired embryo development, and its use has been banned in Britain and other western countries.

Sows normally do not come on heat during lactation although many have a non-fertile heat about 2 days after parturition. No eggs are shed at this heat, and it probably occurs as a result of the residual effect of the oestrogen produced by the placenta. Oestrus usually occurs about 5 to 10 days after weaning and, if the young pigs are removed soon after birth, a fertile heat period may occur as early as 8 to 16 days after farrowing. Under intensive conditions, early weaning is practiced so as to allow maximum frequency of farrowing; synchronization of weaning then provides some synchronization of the post-weaning oestrus.

Treatment with prostaglandin F analogues causes regression of corpora lutea aged 12 days or more, and so will induce abortion; but following such treatment close to term, normal parturition ensues—with very close synchronization, as Fig. 6.7 shows—so that (to some extent) it may be possible to avoid weekend and night time farrowings. Synchronized parturitions also facilitates equalization of litter sizes by exchange of piglets between litters.

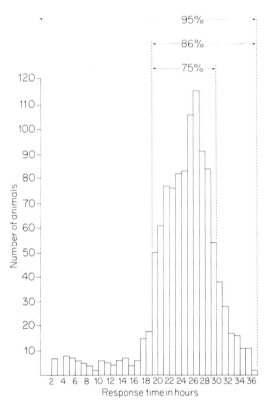

Fig. 6.7 Hourly distribution of farrowing following a morning injection of cloprostenol (an analogue of prostaglandin $F_{2\alpha}$) to 1169 sows, 1, 2 or 3 days before the expected date of parturition. (D. Hammond, ICI Pharmaceutical Division.)

Milk supply and liveweight growth

Variation in weight of pig foetuses increases as pregnancy advances and is greatest in large litters (Pomeroy, 1960). Piglets from large litters are lighter than those from small litters and the position in the uterus where they implant is important (Table 6.3). Where only three embryos were implanted in one horn of the uterus none had any advantage. Where 5 were implanted, those in the middle were significantly lighter at the end of pregnancy, and when 7 were implanted this disadvantage showed up as early as 70 days. The sow differs from the ewe in that underfeeding in late

Table 6.3 Influence of position in the uterus on the weight in grams of the foetuses in the sow. (From Salmon-Legagneur, E. (1968). In *Growth and Development of Mammals*, edited by G. A. Lodge and G. E. Lamming. Butterworth, London.)

Number of foetuses in one horn of uterus	Position in uterus	Stage of pregnancy (days)		
		70	90	110
3	Upper	167	849	950
	Middle	168	728	1013
	Lower	177	732	1153
5	Upper	168	591	1109
	Middle	165	789	882*
	Lower	187	759	1214
7	Upper	178	584	1271
	Middle	140*	472*	967*
	Lower	172	622	1132

* Significantly lighter than litter mates

pregnancy has little effect on size of her young at birth (Lodge, Elsley and McPherson, 1966; Table 6.4).

The mammary glands are formed by a downgrowth of the skin, and when the skin is black the black pigment is carried down to the ducts of the glands and gives the belly a speckled appearance, called 'seedy cut' (Fig. 6.8). As the public is apt to think that the bacon has gone mouldy, the belly has to be cut off before the side is sold and so its presence causes loss to industry. Only traces are present in hog pigs as the udder does not

Table 6.4 Influence of level of feeding of sow on rearing efficiency. (From Salmon-Legagneur, quoted by Rerat, A. (1966). *World Review of Animal Production*, **2**, 39.)

1. *Gestation and parturition*	*Low*		*High*	
Weight at mating (kg)	218		234	
Weight at end of gestation (kg)	260		311	
Weight after parturition (kg)	236		288	
Feed consumption (kg)	217		417	
Number of piglets born alive	11.8		10.1	
Number of stillborn piglets	1.2		1.9	
Average weight of piglets (kg)	1.20		1.24	
Total weight of litter (kg)	15.3		14.2	
2. *Lactation*	*Low*	*High*	*Low*	*High*
Number of sows	8	8	8	8
Quantity of milk produced (kg)	270	315	282	347
Feed consumed (kg)	170	348	165	270
Loss of weight of sows during lactation (kg)	40	4	59	44
Number of piglets at weaning	8.4	8.5	7.5	7.8
Average weight of piglets at weaning (kg)	13.1	16.5	15.3	14.9
Total weight of litter at weaning (kg)	110	140	113	116
3. *Complete cycle*				
Feed consumption (kg)	387	565	582	687
Variation of weight of sow (kg)	−22	+14	−4	+10

develop; the amount in sow pigs is reduced by spaying. About the middle of pregnancy, as the gland develops in preparation for lactation, the pigment vanishes and does not subsequently reappear.

Sow's milk is more concentrated than that of the cow, and contains about 10 per cent fat. It is let down to the young pigs by secretion of oxytocin set free into the blood stream, by the act of sucking which occurs at about hourly intervals. The piglets rub their noses forcefully into the mammary region, which stimulates the release of oxytocin and this causes the ejection of milk, which lasts only a few seconds when the piglets suck vigorously. Each piglet has its own teat. The teats at the fore end of the sow yield more milk than those at the hind end and this accounts for some of the variation in the 3-week weight of young pigs.

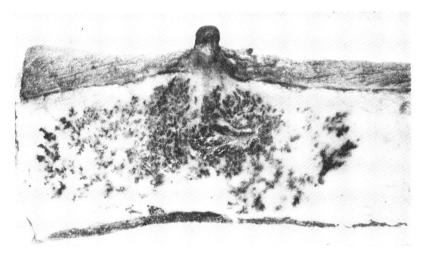

Fig. 6.8 Slice through the belly of a gilt, cut longitudinally through the teat-line, showing black pigment in the ducts of the mammary gland—'seedy cut'. (Cole, L. J., Park, J. S. and Deakin, A. (1933). *University of Wisconsin Agricultural Experimental Station Research Bulletin*, 118.)

Sows average about 300 kg milk in a lactation of 56 days, but there is great individual variation, due to the genetic capacity and age of the sow and to the number of young suckled. As Table 6.4 shows, high-plane feeding of the sow during pregnancy does not contribute very effectively to lactation yield; extra feed is far more profitably supplied during lactation. The lactation curve rises to a peak at about the third week of lactation and then declines (Fig. 6.9), so that supplementary creep feeding is necessary after the third week.

There are now two commonly accepted ways of rearing young pigs— the traditional method of suckling for 6 to 8 weeks with creep feeding from the third week, and the more recent method of early weaning. Piglets can be made to eat from birth; but unless they receive colostrum they have no natural immunity from infections. This absorption of antibodies from the gut into the body occurs in the first day or two of postnatal life; but

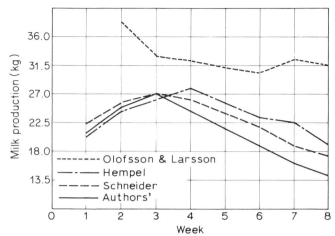

Fig. 6.9 Lactation curves in sows, obtained by weighing the young pigs before and after suckling. Most workers agree that the milk yield begins to fall after the 3rd week. (Bonsma, F. N. and Oosthuizen, P. M. (1935). *South African Journal of Science*, **32**, 360.)

antibodies in the sow's milk help to prevent digestive disturbances after this stage. There is thus an optimum stage for weaning which is presently considered to be at about 20 to 25 days after farrowing. Good litter size and milk production are ensured by use of crossbred sows (see p. 250).

The newborn piglet has a large surface area, relative to its mass, and is poorly insulated; in still air its lower critical temperature is above 30 °C. Consequently it is vulnerable to poor housing. Air movement—draughts—dampness and poorly insulated flooring will cause increased heat loss, so that (see Fig. 10.14) energy is used to maintain body temperature that otherwise would have gone on growth and the deposition of subcutaneous fat (which, by providing insulation, would lower the critical temperature). The critical temperature is in fact lowered by the piglets' habit of huddling together, and so presenting a lower total surface area; for a group, this temperature is effectively about 26 °C at 10 days, falling to below 20 °C at 25 days old (McCracken and Caldwell, 1980). Even so, housing should be above this temperature if the piglets are to be encouraged to seek creep feed rather than remain huddled together. Even if food supply is regulated so that growth rate remains the same, extremes of environmental temperature can affect the nature of the growth (Fig. 6.10).

Differences in birth weight between litters make little difference in weight at weaning—milk supply and creep feed intake are the most important factors. Differences in birth weight within litters, however, can be important because small piglets cannot compete successfully with their larger litter mates. In general, low growth rate before weaning is associated with compensatory growth later on (Lucas, Calder and Smith, 1959) but light weights at 8 weeks are related to excessive fatness and reduced length of carcase and carcase grade at bacon weight (Table 6.5). Hence weights at 3

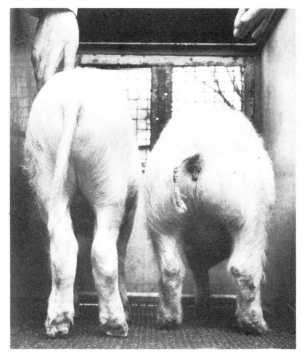

Fig. 6.10 Two litter-mates, the pig on the left exposed to 35°C for seven weeks, and the pig on the right to 5°C; the pig in the cold was fed an amount to keep it at approximately the same body weight as the pig in the warm. (by courtesy of Dr D. L. Ingram and Dr M. E. Weaver). (From Mount, L. E. (1948). *Climatic Physiology of the Pig.* Edward Arnold, London.)

Table 6.5 Percentage of pigs in the bacon-carcase grades related to weight at 8 weeks. (Data from Pig Industry Development Authority, 1962, cited by Lucas, I. A. M. in *Growth and Development of Mammals*, edited by G. A. Lodge and G. E. Lamming. Butterworth, London.)

Bacon carcase grade at 90 kg*	Average weight per pig at 8 weeks (kg)		
	11–13	16–18	20 and over
AA+	55.2	60.3	66.4
AA and A	13.8	13.9	13.4
B+	22.2	19.4	15.9
B, C, F	8.4	5.9	3.7
L	0.4	0.5	0.6
Total %	100.0	100.0	100.0

* Fatness increases from Grade AA to F. AA+ and B+ attain standard for carcase length. L downgraded because of insufficient fat

weeks and at weaning are used by Pig Recording Societies as indices respectively of milk yield of the sow and of growth potential of the pigs.

Remarkable fundamental studies on growth have been made on pigs by McCance and his collaborators. By rearing them at high environmental temperatures, on a very restricted food intake, they were able to keep them to a year old weighing only 4.5 to 5.5 kg. On full feeding, growth followed an S-shaped curve very like that of pigs allowed to grow normally from birth, but the growth curves reached a plateau at about the same age as for normal pigs, but at a lower weight, and they had a greater proportion of fat in their stunted carcases (Lister and McCance, 1967). Tooth eruption was determined by chronological age, while jaw development was more affected by the check to growth.

Market requirements

Pigs were formerly kept for three purposes; to produce either pork (fresh pig meat), bacon or lard. The lard pig was killed at heavy weights, and was effectively a machine kept to convert carbohydrate (maize) into fat. Such a use has long ceased to be economic. Not many years ago the classification, in Great Britain, would have been into pork, bacon or 'manufacturing' pigs. The latter, being above bacon weight, were processed to give sausages, pork pies and other factory products. The present 'heavy pig' is again one which is sold to large-scale processors who can utilize the by-products. Part of its carcase is cured to produce bacon rashers, and part of it is sold as fresh pig meat.

The general requirements for meat production are considered in Chapter 4. In producing pigs for bacon, one consideration is a standard sized animal, of about 90 kg liveweight, because curing is a mass-production operation. Briefly, the main points required are length but not depth of side, 'blockiness' of eye muscle (see Fig. 12.6), fineness of shoulder and even distribution of not too thick subcutaneous fat along the back. Good and poor bacon sides are illustrated in Fig. 6.11. Length of side implies a greater proportion of the side yielding back rashers. Also, the longer the back, the thinner is the layer of a given amount of back fat.

The 'pork' pig is either used to produce quality joints for roasting or— if a rather larger animal—is a 'cutter' used to provide chops etc. for midweek rather than weekend trade. Formerly the pork pig was of about 45 kg liveweight, but the present cutter now nearly reaches bacon weight. The main reason for killing at higher weight is that the older pig requires less-concentrated (and so cheaper) food, and thus is less in competition with poultry and human food needs; also the cost of the weanling pig is a smaller fraction of the final cost. The heavy pig (110 to 120 kg liveweight) produces poorer quality meat, and some skin and fat has to be pared off before sale; but the meat is cheaper to produce.

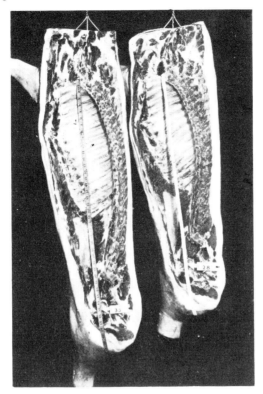

Fig. 6.11   An example of a good bacon carcase (left) compared with a poor one. (Harrington, G. and Pomeroy, R. W. (1955). *Journal of Agricultural Science*, **45**, 431.)

Proportional growth

As the pig grows, its body proportions change. At birth the head and legs are relatively large, and the body is short and shallow; as growth proceeds the body first lengthens and then deepens. In the young pig the lower part of the leg, consisting mainly of bone, is relatively large, while the muscles of the thigh are poorly developed (Fig. 6.12). As the pig grows, the muscles of the back and thigh develop so that the upper part of the leg becomes convex instead of concave, and the valuable loin region develops. These are the differences between good and poorly shaped hams; selection for good shape of ham has been selection for this sort of change in relative growth rates of different parts of the leg. In the improvement of the pig (Fig. 6.13) change of differential growth rates has relatively increased the length, increased the hams and decreased the shoulder and other less valuable parts.

Growth gradients exist between the different tissues in any area of the body. McMeekan followed the development of the Large White from birth to bacon weight; bone develops earlier than muscle, and muscle earlier

Fig. 6.12 The changes in the proportions of the pig as it grows up. In order to compare changes in proportions as distinct from changes in size, all the animals are shown to the same height at the shoulder. The pork breeds such as the Middle White (*right*) pass through the growth changes more quickly than the bacon types such as the Large White (*left*). (*Top*) 3 weeks old; (*second line*) 45 kg (pork weight); (*third line*) 90 kg (bacon weight); (*bottom*) adult. (Hammond, J. (1932). *Journal of the Royal Agricultural Society of England*, **93**, 131.)

than fat (Fig. 6.14). Further, the back fat in the shoulder region develops earlier than over the loin. At bacon weight muscle and fat were both still growing rapidly, but the growth of fat had overtaken that of muscle.

Breeds differ in the rate and extent of these changes in body proportions (see Fig. 6.12). The early-maturing pork types, such as the Middle White, pass through these changes more quickly, and to a further extent, than the later-maturing bacon type such as the Large White. At 45 kg liveweight the Middle White has good hams and is well finished, while the Large White still has a high proportion of bone (Fig. 6.15). However, at bacon

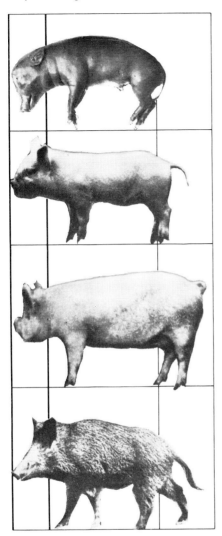

Fig. 6.13 How the anatomy of the pig has been changed to meet market requirements. Each animal is shown at same head size. As an improved breed like the Middle White grows up, the proportion of loin to head and neck increases greatly; but an unimproved type such as the Wild Boar grows up without much change in body proportions. (*Top*) Foetus of 2 months; (*second line*) Middle White, 3 weeks old, 7 kg; (*third line*) Middle White, 15 weeks old, 45 kg; (*bottom*) Wild Boar, adult, about 135 kg. (Hammond, J. (1933–4). *The Pig Breeders' Annual, London.*)

weight development in the Middle White has gone too far: the side is too deep, and too fat. In other words, the early-maturing type has much the same body proportions at 45 kg as the later-maturing type has at 90 kg liveweight. To obtain the best results in marketing the short deep type should be slaughtered at lower weight than the longer pigs.

When the type of production changes, as has occurred in the U.S.A. with the fall in the price of lard, breed type can be changed by selection. The lard type of pig is a small-framed animal with a great thickness of fat. By selection for increased size and for later maturity the Poland–China has

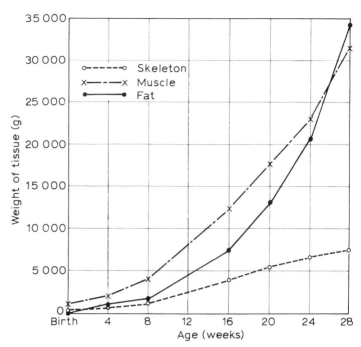

Fig. 6.14 Changes in the composition of the carcase of the Large White pig as it grows up. (McMeekan, C. P. (1940). *Journal of Agricultural Science*, **30**, 276.)

been changed (Fig. 6.16) from a lard type to one nearer a modern bacon type.

The pig differs from other species of farm animal in that there is little sex difference in body size. While (at equal body weight) the castrated male, in sheep or cattle, has a higher proportion of muscle and less fat than the female, in pigs the position is reversed. At bacon weight the gilt has better development of 'eye muscle' (longissimus dorsi) in the loin, and less fat over it, than has the hog pig. As in other species, the entire male has far less fat and therefore has a better apparent food-conversion efficiency (see p. 33). The entire male grows faster and produces a leaner carcase. Young boars can be penned with sows up to marketing age—which is before puberty—and there is no taint in their meat. Castration of boars therefore appears to be unnecessary and wasteful.

Plane of nutrition and development

By controlling the rate of growth, and hence the rate of deposition of fat relative to the growth of bone and muscle, it is possible to alter the proportion of bone, muscle and fat in the carcase. McMeekan controlled the plane of nutrition in different stages of growth and grew pigs along

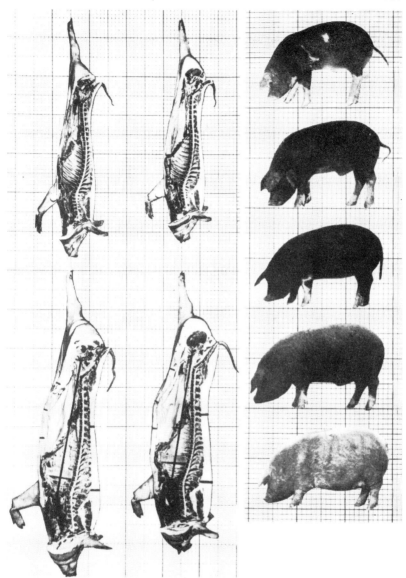

Fig. 6.15 (left) Carcases of Middle White (*right*) and Large White (*left*) at 45 kg liveweight for pork (*above*) and at 90 kg liveweight for bacon (*below*). Compare with the live animals shown in Fig. 6.12. (Hammond, J. (1932). *Journal of the Royal Agricultural Society of England*, **93**, 131.)

Fig. 6.16 (right) Changes in body proportions which have taken place in the Poland-China pig as the demand for lard decreased. (*From bottom*) 1895–1912; 1913; 1915; 1917; 1923. (Hammond, J. (1932). *Journal of the Royal Agricultural Society of England*, **93**, 131.)

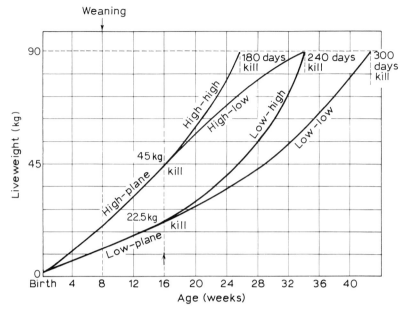

Fig. 6.17 Plan of McMeekan's experiment to determine whether the composition and conformation of the pig can be changed by altering the shape of its growth curve. (McMeekan, C. P. and Hammond, J. (1939). *Journal of the Ministry of Agriculture, London*, **46**, 238.)

predetermined growth curves of very different shape (Fig. 6.17) so as to produce extreme difference of size at 16 weeks of age. A comparison was then made of the proportions and compositions of some of them, while the rest were either continued on the same planes until they reached 90 kg liveweight, or were interchanged. Thus there were four groups killed at bacon weight, two of them of the same age, but grown along differently shaped curves.

At 16 weeks, as Fig. 6.18 and Table 6.6 show, the smaller low-plane animals are relatively immature and 'unimproved' in their proportions and

Table 6.6 Per cent composition at 16 weeks of pigs reared on a high or low plane of nutrition expressed on a total-weight basis or on a fat-free basis. (Data from McMeekan, 1940–41.)

Percentage	Total weight basis		Fat free basis	
	Low plane	High plane	Low plane	High plane
Fat	10.3	30.7	–	–
Bone	19.2	12.7	21.4	18.4
Muscle	53.1	45.8	59.3	66.0
Skin, tendons, etc.	14.3	8.7	15.9	12.6
Dissection losses	3.1	2.1	3.4	3.0
Total	100.0	100.0	100.0	100.0

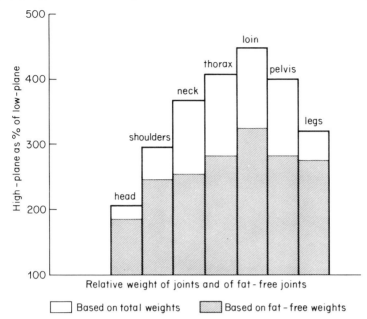

Relative weight of joints and of fat - free joints

☐ Based on total weights   ▨ Based on fat - free weights

Fig. 6.18 Differences, at 16 weeks old, in the body proportions of pigs fed on high and low planes of nutrition. The weight of the joint in the high-plane pigs is shown as a percentage of the weight of the joint in low-plane pigs on either a total-weight or fat-free basis. (From McMeekan, 1940–41, modified by inclusion of fat-free data.)

composition. Relatively the loin is much more stunted than the head. The amount of fat contributes considerably to the differences in the weights of different parts of the carcase, but on a fat-free basis the effect is still very marked. An undernourished animal cannot be expected to express its full potential for rapid growth rate, nor will it show its capacity for developmental changes in body proportions.

Some of the effects of shape of growth curve, to 90 kg liveweight, are illustrated in Figs 6.19 to 6.21. Bone is the major tissue least affected by under nutrition; the older animals have greater growth of bone, and are longer (the very old L–L animal in Fig. 6.19 appears short because of its great length of limb); Table 6.7 shows the proportions, by weight, of the carcase components.

Comparing the H–H and L–L animals, the older one has poorer hams, but better appearance at the last rib (Fig. 6.21). On a high plane, with unrestricted growth of fat and muscle, there is more fat, and it has been laid down on a shorter frame. Because of its much greater age, the food cost of the L–L animal is greater, although its energy content is less: more will have been used for maintenance.

It is the comparison between the animals (L–H and H–L) of the same age which is of greater interest. The L–H pig is shorter and fatter than the one fed well at first and checked later. Though in the earlier stages fat was

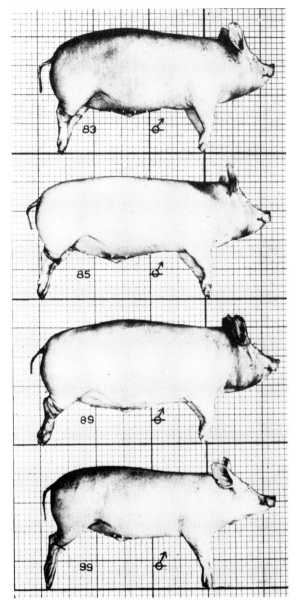

Fig. 6.19 Pigs of bacon weight (90 kg) grown at different rates (Fig. 6.17); all shown to same height at shoulder. For details see Fig. 6.21. (McMeekan, C. P. and Hammond, J. (1939). *Journal of the Ministry of Agriculture, London,* **46**, 238.)

Table 6.7   Per cent composition at 90 kg of pigs reared on High–High, High–Low, Low–High or Low–Low nutritional regime. (Data of McMeekan, 1940–41.)

Percentages	High–High	High–Low	Low–High	Low–Low
1. Total weight basis				
Fat	36.9	28.7	40.1	26.6
Bone	11.1	11.6	10.4	12.6
Muscle	41.7	48.1	38.8	49.3
Skin, tendon, etc.	8.5	9.9	8.5	10.5
Dissection losses	1.8	1.7	2.2	1.0
Total	100.0	100.0	100.0	100.0
2. Fat free basis				
Bone	17.7	16.3	17.4	17.2
Muscle	66.1	67.6	64.9	67.2
Skin, tendon, etc.	13.5	13.9	14.2	14.3
Dissection losses	2.7	2.2	3.5	1.3
Total	100.0	100.0	100.0	100.0

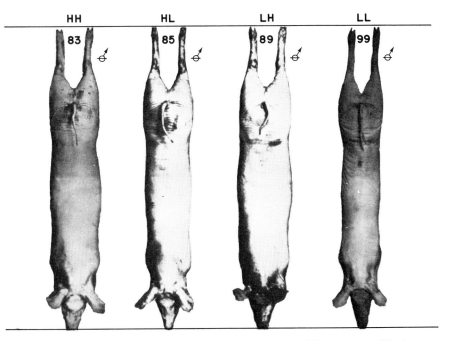

Fig. 6.20   Carcases of pigs of bacon weight grown at different rates. All pigs shown to the same length. For details see Fig. 6.21. (McMeekan, C. P. and Hammond, J. (1939). *Journal of the Ministry of Agriculture, London,* **46**, 238.)

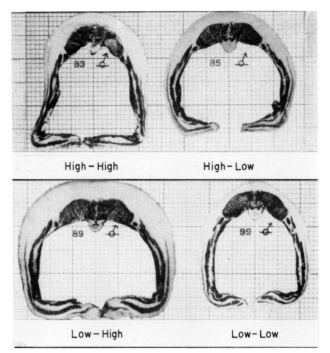

Fig. 6.21 Cut through the last rib of bacon pigs grown at different rates. All photographs reduced to the same 'eye muscle' length.

83 H-H   Quickly grown throughout. 168 days old.

85 H-L   Quickly grown to 16 weeks, afterwards grown slowly. 196 days old.

89 L-H   Slowly grown to 16 weeks, afterwards grown quickly. 196 days old.

99 L-L   Slowly grown throughout. 315 days old.

(McMeekan, C. P. and Hammond, J. (1939). *Journal of the Ministry of Agriculture, London*, **46**, 238.)

much more affected than bone, growth in length of bone was checked. When, after 112 days, the L–H animals were fully fed there was much more rapid growth of fat than of muscle and bone. In the H–L pigs the effect of the later restricted food intake was mainly to hold back fat deposition and so to produce a larger and leaner carcase, much more desirable than that of the L–H animals. Less severe restriction in the earlier stages of growth would be expected to have much less marked later effects.

Since McMeekan's experiments were carried out, pigs have been improved for bacon production (see Chapter 12) by breeding for greater growth potential of bone and muscle and later maturity (in the sense of fat deposition). Less restriction of feed intake, and more rapid growth, is thus compatible with good bacon carcase quality. It has been pointed out that larger adult size entails later sexual maturity and higher maintenance requirement of the brood sows.

The alternative method (H–L pattern feeding) of obtaining good carcases, though less efficient in the amount of food consumed and time taken to reach market weight, is in practice obtainable by feeding less-concentrated and cheaper feed to appetite, rather than by restricting feed supply; on such a diet the pig is less directly in competition with man.

References

DHINDSA, D. S. and DZUIK, P. J. (1968). Effect on pregnancy in the pig after killing embryos or foetuses in one uterine horn in early gestation. *Journal of Animal Science*, **27**, 122, 668.

HEITMAN, H. and COLE, H. H. (1956). Further studies on the induction of oestrus in lactating sows with equine gonadotrophin. *Journal of Animal Science*, **15**, 970.

HUNTER, R. F. H. (1966). The effect of superovulation on fertilization and embryonic survival in the pig. *Animal Production*, **8**, 457.

LISTER, D. and MCCANCE, R. A. (1967). Severe undernutrition in growing and adult animals: 17. The ultimate results of rehabilitation in pigs. *British Journal of Nutrition*, **21**, 787.

LODGE, G. A.,ELSLEY, F. W. H.and MCPHERSON, R. M. (1966). The effects of level of feeding of sows during pregnancy. I and II. *Animal Production*, **8**, 29 and 499.

LODGE, G. A. and MCPHERSON, R. M. (1961). Level of feeding during early life and the subsequent reproductive performance of sows. *Animal Production*, **3**, 19.

LUCAS, I. A. M., CALDER, A. F. C. and SMITH, H. (1959). The early weaning of pigs. VI. *Journal of Agricultural Science*, **53**, 136.

MCCRACKEN, K. J. and CALDWELL, B. J. (1980). Studies on diurnal variation of heat production and the effective lower critical temperature of early weaned pigs under commercial conditions of feeding and management. *British Journal of Nutrition*, **43**, 321.

POMEROY, R. W. (1960). Infertility and neonatal mortality in the sow. I to IV. *Journal of Agricultural Science*, **54**, 1, 18, 31, 57.

SIGNORET, J. P. (1972). The mating behaviour of the sow. In *Pig Production* (D. J. A. Cole, Ed.). Pennsylvania State University Press.

ZIMMERMAN, D. R., SPIES, H. G., SELF, H. L. and CASIDA, L. E. (1960). Ovulation rate in swine as affected by increased energy intake just prior to ovulation. *Journal of Animal Science*, **19**, 295.

Further reading

COLE, D. J. A. (Ed.) (1972). *Pig Production. Proceedings 18th Easter School Nottingham.* Pennsylvania State University Press.

DZUIK, P. J. (1977). Reproduction in Pigs. In *Reproduction in Domestic Animals*, third edition (H. H. Cole and P. T. Cupps, Eds). Academic Press, London and New York.

LARSSON, K. (1978). Current research on the deep freezing of boar semen. *World Review of Animal Production*, **9**, 45.

7 Poultry

The breeding season

Wild birds of the temperate zone have, in general, a restricted breeding season, regulated primarily by length of day. The influence of photoperiod and of other factors is co-ordinated in the hypothalamic nerve centres which, through the pituitary, regulate the sex functions and body metabolism generally.

The season begins in spring, as the days get longer, but ends while they are still long (and sometimes still increasing) owing to a phenomenon known as refractoriness, which is a delayed response to long day-length. At the end of the breeding season there is a moult, and the inactivity of the

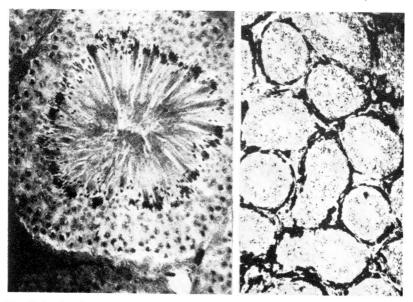

Fig. 7.1 Sections through the testes of starlings during the winter non-breeding season. (*Left*) Given electric light in front of cage: tubules large and spermatozoa being formed. (*Right*) Normal, not lighted: tubules small and inactive. (Bissonette, T. H. and Wadlund, A. P. R. (1932). *Journal of Experimental Biology*, **9**, 339.)

gonads is also marked by the cessation of singing. With the ending of refractoriness there is a resumption of bird-song (stimulated by androgen from the ovary or testis), though day-length may still be decreasing. There is a further moult, and accumulation of body fat, in the autumn.

The male, if it is not still refractory, will respond to extra illumination in autumn or winter by the premature enlargement of the testis (Fig. 7.1). Full development of the ovary, and egg laying, generally require more complex stimuli, as well as long photoperiod. Furthermore, egg laying does not normally continue all through the breeding season. When the appropriate number of eggs are in the nest, the pattern of pituitary activity is changed, laying ceases, and incubation begins—though laying will continue if eggs are removed as they are laid.

The chicken differs from wild birds in several respects. It shares with related wild birds, (e.g. pheasant) the capacity to develop the ovary and to lay when kept in isolation. Other differences are probably a result of modifications by selection under domestication. There is a seasonal variation in rate of egg laying under natural lighting conditions (Fig. 7.2), but laying can occur at all times of the year. It may be noted from this figure that the seasonal alteration in intensity of lay matches the amplitude of the changes of day length at different latitudes: also that the pattern of rate of lay anticipates that of day-length change.

There is no clear-cut phenomenon of refractoriness, but fowls kept on long days indefinitely do not continue to lay well, while fowls kept on short day-length, though they do not lay well at first, will increase their rate of laying.

Egg laying

The bird has only one functional ovary and oviduct, though destruction of the ovary may result in masculine development of the rudimentary right gonad, and an apparent 'change of sex'. The protein and lipids of the egg yolk are synthesized in the liver, under the influence of oestrogen, and are transported to the ovary by the blood stream. At normal ovulation a single egg yolk is shed from the ovary and is caught in the funnel at the top of the oviduct. In the next four hours the upper part of the oviduct deposits the egg white, and the succeeding portion the shell membranes. In the shell gland, or uterus, the egg then spends 20 hours or more. During this period the white takes up water, so that the egg 'plumps', and the mineral for the shell is gradually withdrawn from the blood and deposited onto the membrane. At laying, the egg passes through the vagina and is sealed with a layer of mucin, which protects it against bacterial invasion. If the muscle of the oviduct is made more excitable, as can be done by mechanical muscle irritation or by administration of drugs, the egg is laid prematurely as a soft-shelled egg.

A hen lays in clutches, which may consist of one to seven or more eggs, followed by a pause of one or more days before a second clutch is begun. With increased productivity, clutch size increases and the intervals between

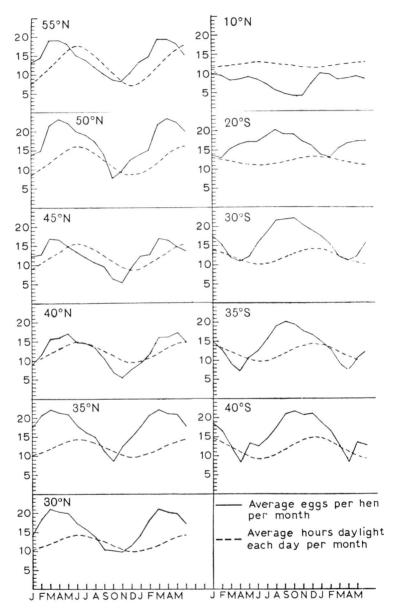

Fig. 7.2   Seasonal egg production in fowls in different latitudes compared with the seasonal daylight curves. (Whetham, E. O. (1933). *Journal of Agricultural Science*, **23**, 383.)

clutches decrease (Fig. 7.3). Most hens ovulate in the morning and very few do so after 3 p.m. The egg spends upwards of 24 hours in the oviduct and the next ovulation of the clutch occurs $\frac{1}{2}$ to 1 hour after laying. Hence the hen tends to ovulate later each day, and eventually is due to ovulate late in the afternoon. She then holds back, misses a day or more, and then begins a fresh clutch.

Nalbandov has increased expected clutch length by injection of gonadotrophin, and there is little doubt that clutch length depends on the level of pituitary activity. Increased productivity can be obtained by artificial lighting to supplement short day-length in winter. Treatment with gonadotrophins can cause ovulation of up to 7 yolks at a time (Fraps and Riley, 1942) but in such experiments only one yolk enters the oviduct. Double-yolked eggs, however, are due to natural twin ovulations.

	1	2	3	4	5	6	7	8	9	10	11	12	13	14	15	16	17	18	19	20	21	22	23	24	25	26	27	28	29	30	31	Total
1914																																
May	1		1	1	1		1	1	1	1	1	1		1	1	B				O				1	1		1	1	1	1		18
June	1	1	1	B		O							1	1			1	1						1	1	1	1	1	1	1	x	4
1968 M																																
May	1		1	1	1		1		1	1	1		1	1	1			1	1		1	1	1		1		1	1		1	1	21
June	1	1			1	1		1	1	1		1	1			1	1	1		1		1	1	1			1	1		1	x	19
1968 G																																
May	1	1	1	1	1	1	1		1	1	1	1	1	1	1		1	1	1	1	1	1	1		1	1	1	1	1	1	1	28
June		1	1	1	1	1	1		1	1	1	1	1	1		1	1	1	1	1	1		1	1	1	1	1	1	1		x	25

Fig. 7.3   Egg records of a hen in 1914 (Pearl, R. J. *Animal Behaviour*, **4**, 266) compared with records of two hens in 1968, one a mediocre and the other a good layer, showing the improvement in performance due to longer clutches and elimination of broodiness. **B** denotes the beginning and **O** the end of the brooding periods.

A prolific hen lays clutches of five or more eggs, with pauses of no more than a day, while clutches of 50 to 100 eggs are not uncommon and records are known of 365 eggs in a year. Hens laying large clutches achieve this by reducing the intervals between ovulations to little more than 24 hours and by reducing the time the egg spends in the shell gland.

During broodiness—while the instinct to incubate operates—egg laying is suppressed, and there is an interval after broodiness ends before laying is resumed (Fig. 7.3). Maintenance of broodiness in the bird appears to be closely analogous to maintenance of lactation in the mammal, and to depend on adequate breast stimulation. Tendency to go broody has been much reduced by genetic selection.

The age at first ovulation, and level of pituitary gonadotrophic activity, is determined, as it is in the sheep, by interaction between age, day-length and nutrition. As in mammals generally, reproductive activity begins well before growth is completed, and earlier under good nutritive conditions (Table 7.1). Oestrogen is the main hormone produced by the developing ovary, but androgen is responsible for the growth and reddening of comb

Table 7.1 Food supply in relation to maturity in White Wyandotte pullets. (From Prentice, J. H., Baskett, R. G. and Robertson, G. S. (1930). *Proceedings of the 4th World's Poultry Congress, London.* No. **37**, 224.)

Ration	Age in days when first egg laid	Average weight in gm when first egg laid
Basal	186	1715
Basal + minerals	146	1726
Basal + milk	137	1830
Basal + soya bean + minerals	141	1809
Intensive ration	135	1800

and wattles, and both androgen and progesterone synergize with oestrogen in the development of the oviduct.

The age at first ovulation, under natural lighting conditions, depends upon the date of hatching (Fig. 7.4). It is not however necessarily advantageous to bring a bird into lay at as young an age as possible. Laying and body growth are in competition; the first eggs laid by the young bird are small and not so readily saleable as the eggs laid when the bird is better

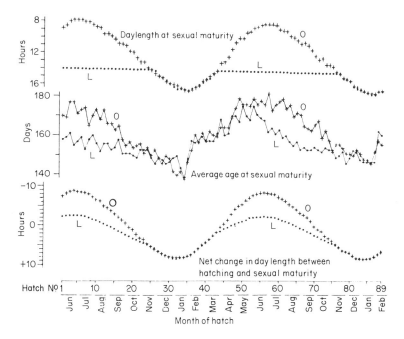

Fig. 7.4 Average age at first egg (centre figure) for groups of pullets hatched at weekly intervals throughout 21 months and raised with natural daylight only **(O)** or with supplementary artificial light from the age of 16 weeks onwards **(L)**. The upper figure shows the day length at the time of average first egg for each group; the lower figure shows the change in day-length (day-length at maturity minus day-length at hatching) experienced by each group. (Morris, T. R. and Fox, S. (1958). *Nature, London,* **181**, 1453.)

grown. Manipulation of day length exposure can delay laying until the eggs will be larger.

Withdrawal of food causes a cessation of laying, but pituitary injections to starved birds can induce egg laying. On a diet deficient in calcium laying also soon stops, but it will continue for a long while if the bird is caused to lay soft-shelled eggs (Gilbert, 1969).

Hens lay most intensively during the first year of life. Although some hens will continue to lay during the first moult, many will not. The general moult of body feathers normally occurs some 16 months after hatching, when the fowl has been in production for 10 or 11 months. The duration of laying appears to be determined partly by genetic factors and partly by the environment. Birds kept on long day can be thrown into a moult by transfer to short days, just as can mammals: such a change can also suppress egg laying. Moulting in birds has been produced with thyroid hormone treatment, but this is not necessarily incompatible with the basic mechanism being similar to that operating in mammals (page 25).

Fertility and artificial insemination

Fertility depends very largely on the number and viability of the sperm produced by the cock, or used in artificial insemination, and also on the time relationships between mating and insemination and laying.

At mating the cloaca is everted and the sperm are deposited in the vagina of the hen. It takes up to two or three hours for the first of them to appear at the top of the oviduct, where the egg is fertilized when it is liberated from the ovary. Consequently the first fertile egg is not usually laid until 30 hours or more after mating, due to the time taken for the egg to pass down the tract (page 173). Fertilization takes place in the infundibulum or funnel of the oviduct within 15 minutes of ovulation and no egg which has started on its way can be fertilized. Although fertile eggs may be laid as soon as 30 hours after mating, full fertility of the eggs laid is not attained until the second day. The duration of the period of maximum fertility varies with the activity of the cock and the quality of the sperm, but in practice it is usually taken as 2 to 6 days. The time taken to attain the maximum percentage of fertile eggs depends on the number of cocks introduced to the flock (Fig. 7.5).

Sperm may retain their fertilizing capacity for as long as 32 days but the usual range is 11 to 14 days. However, when the cock is removed from the pen a drop in fertility is apparent by the 6th day and by the 10th day only 50 per cent of the eggs are fertile. Hence an abundance of active sperma-tozoa in the tract is essential for maximum fertility.

The increased use of cage housing and the concentration of breeding in the hands of a few large organizations has stimulated interest in artificial insemination, techniques for which are described by Lake (1962). Semen is collected by massaging the ejaculatory ducts (Fig. 7.6) through the ab-dominal wall. The volume varies from 0.25 ml to more than 1 ml, and density is similarly variable. Semen is diluted, using one of a number of

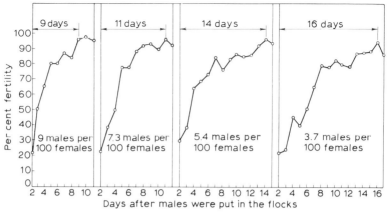

Fig. 7.5 The relation of the number of males used in New Hampshire breeder flocks to the time required for the maximum level of fertility to become established. (Parker, J. E. and Bernier, P. E. (1950). *Poultry Science*, **29**, 377.)

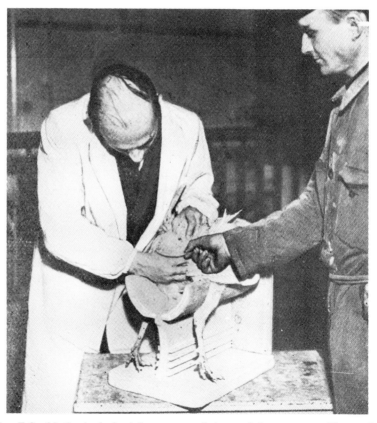

Fig. 7.6 Method of obtaining sperm of the cock by massage. The cock is controlled by placing his head through a hole in the stand and his legs through slots. (Griffini, G. (1938). *Fecondazione Artificiale degli Animali Domestici, Milano*.)

diluents, and is used as rapidly as possible. A dose of 0.1 ml diluted semen (dilution 1:1 to 1:3) containing not less than 80 to 100 million sperm is placed into the cloaca (Fig. 7.7) every seven days. No completely satisfactory method of storing or freezing semen has been devised despite the early promise of the work of Polge, who obtained 54 per cent fertility and 71 per cent hatchability following freezing of fowl sperm to $-79\,°C$.

A good cock will produce semen containing up to $5\frac{1}{2}$ million sperm per cubic millimetre and is said to be able to tread hens up to 20 to 40 times a day but, as shown in Fig. 7.8, the concentration of sperm falls rapidly long

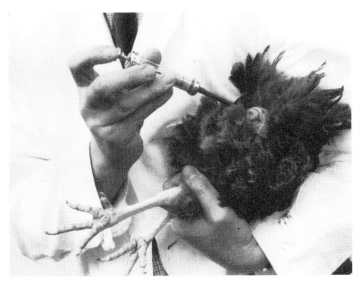

Fig. 7.7   A one-man method of everting the vagina and inseminating a hen. (Lake, P. E. (1962). In *The Semen of Animals and Artificial Insemination*. (J. P. Maule, Ed.) Commonwealth Agricultural Bureaux, Farnham Royal.)

before this frequency of ejaculation is attained. Such a cock, however, can be used to inseminate 50 to 100 hens per week, with twice-weekly collections. Generally, about 90 per cent of eggs are fertilized and about 80 per cent of fertilized eggs hatch. The reasons why fertilized eggs fail to hatch fall into two groups, namely internal factors, i.e. the quality of the eggs, and external factors, i.e. incubation environment.

(*a*) *Internal factors*   At least three types of embryonic development may take place (Fraps, 1955). First, the egg may be fertilized at the correct time by a normal sperm, and fertilization is followed by normal development. Second, a normal egg may be fertilized by a sperm of low viability—such as one which has been in the oviduct a long time—so that development is started but the embryo dies. Third, there may be development of an unfertilized egg (parthenogenesis). Usually such embryos die early, but many turkey chicks have developed up to the 27th day in the egg and a few up to adult life. The best hatches are obtained from eggs laid in the middle of the day and McNally and Byerly found that eggs laid during this period had a particularly high degree of embryonic development; the more rapidly the embryo begins to grow the better the chances that it will complete the process. Differences exist in the time taken for the egg to traverse the tract and this may account for differences in the stage of development of eggs at lay.

(b) *External factors*   If the egg cools to less than 20 °C following laying, its development ceases and it will only recommence when the temperature is raised. The longer the egg is kept before incubation the lower is the percentage hatchability. The optimum temperature for development is

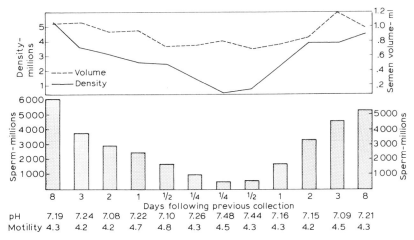

Fig. 7.8   Effect of frequency of collection on characteristics of avian semen. (Parker, J. E., McKenzie, F. F. and Kempster, H. L. (1942). *Research Bulletin of the Missouri Agricultural Experiment Station*, No. 347.)

37.5 °C to 40 °C and the conditions of temperature, humidity and air flow (gaseous exchange) are important. Although death of the embryo may occur at any stage of incubation, there are two main critical periods (Fig. 7.9), at the 4th and 19th day. The mortality is much greater at the 19th day, however, than at the 4th day, and is exaggerated by defects in the incubating environment. Both peaks are connected with the exchange of gases between the eggs and their environment and the accumulation of lactic acid. The second peak coincides with a critical stage of the life of the embryo when its respiration changes from the chorio-allantoic to the pulmonary type.

Growth of the chick

A newly hatched chicken weighs about 60 per cent of the original weight of the egg, and there is little evidence that this weight influences the ultimate weight of the bird. During incubation the chick makes maximum growth during the last few days but the rate of growth, expressed as a proportion of its size at any one time, is linear (Fig. 7.10(a)). Following hatching the growth rate slows down for a few days and then accelerates (Fig. 7.10(b)).

Under extensive conditions the rate of growth of the chick after hatching is much less than that of a mammal such as the rabbit (Fig. 7.11). Murray attributed this to differences in postnatal nutrition for, whereas the rabbit is supplied with milk, a high-protein, high-energy feed, the chick has a sudden change to low-protein bulky feed. There is no inherent reason why birds should not make rapid early growth after hatching, as shown by the

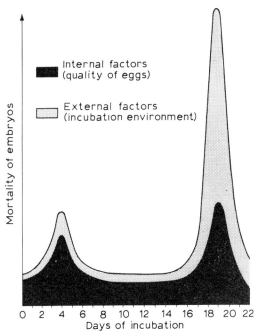

Fig. 7.9 Frequency distribution of embryonic mortality in incubated eggs. (Romanoff, A. L. (1931). *Cornell University Extension Bulletin*, 205.)

pigeon, in which the young are fed by the mother on milk from the crop gland and on semidigested food from the crop; Kaufman in 1929 showed the growth rate of the pigeon in the first month to be over 300 per cent compared with only 160 per cent in the chick of that time. Using modern poultry starters, highly palatable with a high-energy and protein content, coupled with modern systems of housing to reduce heat loss and activity, better growth rates than in Kaufman's pigeons are attained nowadays even with laying strains of birds. The growth rate of the modern broiler is three times as great (Fig. 7.11). Although the adult modern layer is no larger than that of 60 years ago, controlled feeding gives it a different growth curve. Improved knowledge of nutrition enables early rapid growth, so

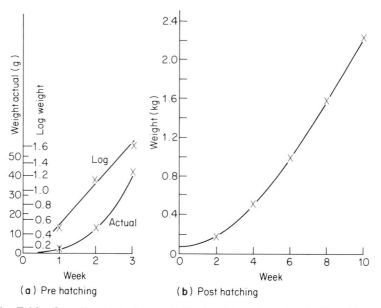

Fig. 7.10 Growth rates before and after hatching of modern broiler chickens. (Hill, 1962 — see Table 7.2.)

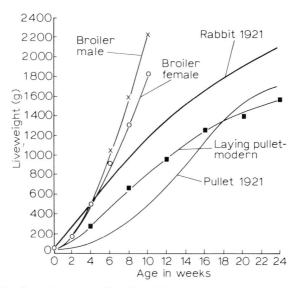

Fig. 7.11 Growth curves of broiler chickens up to marketing age (10 weeks) compared with curves for laying pullets of today and 50 years ago. (Data from Murray, 1921; Hill, 1962 — see Table 7.2.)

that the ages of maximum growth rate and of maturity have been brought forward.

The modern broiler increases its weight more than ten-fold in the first month, trebles it in the next month, and is ready for market at 7 or 8 weeks of age. Its growth rate has not yet reached its maximum (Fig. 7.12) but its efficiency of conversion of feed is falling (Table 7.2) due to the higher proportion of fat being deposited.

Despite this rapid growth and high efficiency of feed utilization, the milk-fed mammal has an advantage very early in life, but it is soon lost (Fig. 7.12). In the chicken and turkey, the male grows at a faster rate and

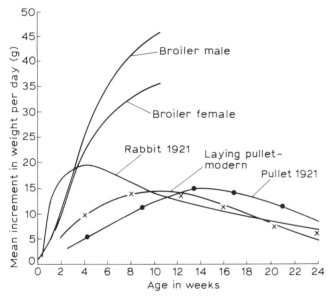

Fig. 7.12 Growth-rate curves showing increase in grams per day for modern broiler chickens compared with laying pullets of today and 60 years ago. The growth-rate curve of a mammal (rabbit) is shown for comparison. (Data from Murray, 1921; Hill, 1962—see Table 7.2.)

for a longer time than does the female. This difference, unlike many secondary sexual characters, remains after castration, spaying, or transplantation of the gonads between the sexes (Fig. 7.13). Table 7.2 shows that between 16 and 28 weeks of age the growing turkey hen requires 9 kg of feed for each kg of gain, as compared with $5\frac{1}{2}$ kg for the cock. This shows her relative earliness of maturity and the slowing down of muscular growth and the increase in high-energy fat.

With the development of a specialized poultry meat industry there is relatively little interest today in caponizing and hormone implants. Caponizing lessens the colour of the flesh, and it also encourages quicker fattening. This effect can also be produced by implanting tablets of the synthetic oestrogen, stilboestrol, under the skin of the neck.

Fig. 7.13 Sexual differences in size in Brown Leghorn fowls. Although the gonads affect the plumage and comb size they have no effect on body size. **1.** Normal cock. **2.** Normal hen. **3.** Castrated cock. **4.** Castrated hen. **5.** Castrated cock with ovaries transplanted. **6.** Castrated hen, with testes transplanted. (Zawadowsky, M. M. (1931).). *Dynamics of Development of the Organism.* Moscow.)

Table 7.2 Representative data illustrating growth rates and feed requirements of poultry. (From Hill, F. W. (1962). In *Introduction to Livestock Production*, edited by H. H. Cole. Freeman, San Francisco.)

Type of bird	Age (weeks)	Average weight (kg)	Cumulative feed consumed (kg)	Feed per kg gain Overall (kg)	During period (kg)
Broiler chickens					
	2	0.2	0.3	1.5	–
Males	6	1.0	2.0	2.0	2.1
	10	2.2	5.4	2.4	2.8
	2	0.2	0.3	1.5	–
Females	6	0.9	1.8	2.0	2.1
	10	1.8	4.5	2.5	3.0
Turkeys					
	4	0.8	1.0	1.2	–
Males	16	7.0	18.8	2.7	2.9
	28	13.4	53.6	4.0	5.5
	4	0.6	0.8	1.3	–
Females	16	5.0	14.6	3.0	3.2
	28	7.8	40.0	5.1	9.0

Changes in body proportions

The body proportions of the chick change as it grows up in much the same way as do those of mammals. The changes in the proportion of fat and muscle to bone take place more rapidly in pullets than in cocks, and the pullet has a greater capacity for the storage of fat than has the cock. This difference becomes most marked at the onset of sexual maturity. Wilson (1952) investigated the changes in proportions of the body in fowls reared on high and low planes of nutrition up to 24 weeks old, and in two groups in which the plane of nutrition was switched over at 10 weeks, with results similar to those reported in sheep and pigs. The bird shares with the pig a remarkable recuperative capacity following a long period of under-nutrition (McCance, 1960).

References

FRAPS, R. M. (1955). Egg production and fertility in poultry. In *Progress in the Physiology of Farm Animals* (J. Hammond, Ed.), chapter 15. Butterworth, London.

FRAPS, R. M. and RILEY, G. M. (1942). Hormone induced ovulation in domestic fowl. *Proceedings of the Society for Experimental Biology and Medicine*, **49**, 253.

GILBERT, A. B. (1969). The effect of a foreign object in the shell gland on egg production of hens on a calcium-deficient diet. *British Poultry Science*, **10**, 83.

McCANCE, R. A. (1960). Severe undernutrition in growing and adult animals. *British Journal of Nutrition*, **14**, 59.

WILSON, P. N. (1952, 1954). Growth analysis of the domestic fowl. I, II and III. *Journal of Agricultural Science*, **42**, 369; **44**, 67; **45**, 110.

Further reading

BELL, D. J. and FREEMAN, B. M. (1971). *Physiology and Biochemistry of the Domestic Fowl*. Academic Press, New York and London.

MURTON, R. K. and WESTWOOD, N. J. (1977). *Avian Breeding Cycles*. Clarendon Press, Oxford.

NALBANDOV, A. V. (1976). *Reproductive Physiology*, third edition. Freeman, San Francisco.

Part 2

Genetics and Breeding

8 General considerations

Animal and plant breeding

Although the basic principles underlying both plant and animal breeding are the same, their application to plants and animals has been different for several reasons. First, the numbers of individuals—the population sizes—which are available and can be handled conveniently in breeding and selection programmes are very much greater for plants than for animals. Second, the time that must elapse between one generation and another, the generation interval, is usually shorter in plants than in animals. Third, most individual plants can be fertilized by themselves or propagated vegetatively and so pure lines of plants can be obtained more easily than pure lines of animals.

These differences have had some important consequences for the methods and achievements of plant and animal breeders. For more than fifty years plant breeding has been done by geneticists in national and commercial organizations who select, and organize the multiplication of, improved strains of plants for sale to farmers. The influence on animal improvement of organizations employing geneticists has been a more recent development; the role of the individual farmer-breeder has remained important. However, most farmer-breeders of livestock now work in cooperation with each other or with national or commercial organizations. This allows selection in animal populations which are often as large as those available to plant breeders.

Another consequence of the differences between plant and animal breeding is that the rate of change per year in performance which can be achieved by selection is greater for plants than for animals. The development of new techniques for stimulating earlier and more prolific reproduction is increasing the rate of change in both plant and animal breeding.

Domestication

Domestication is the process of developing strains of plants and animals which can be bred, handled and selected by man. Our domestic species and breeds of livestock originated, as did the cultivated varieties of plants, in different early centres of civilization and spread outward as

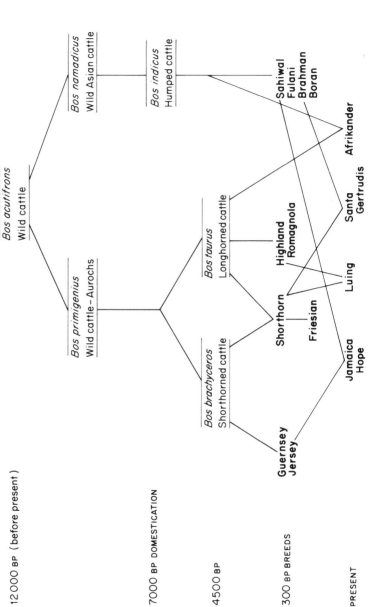

Fig. 8.1 A schematic development of modern cattle breeds by domestication from wild cattle.

the various civilizations extended throughout the world. Reindeer, dogs, sheep, goats, pigs, poultry and cattle were domesticated, probably in that order, over the period from 14 000 to 5 000 years ago. Much of the evidence for domestication comes from sites in northern Eurasia in the area between the rivers Euphrates and Tigris in Iraq and from Turkey, Syria and Jordan. From 5000 to 4000 years ago domesticated horses appeared in Asia and donkeys in north Africa. Domestication of the llama and guinea pig in Peru, the camel in central Asia and the turkey in Mexico probably also occurred at that time. Attempts at domestication have not always been successful and it is known that efforts to domesticate the jackal and the hyena failed. New efforts to domesticate previously wild species such as red deer in Scotland and eland in Kenya are making progress.

The reasons why man domesticated animals are difficult to determine, though Zeuner (1963) has indicated a biological interpretation and Isaac (1970) has suggested that domestication may be associated with the origins of religion.

It is almost certain that domestication of a species did not occur at a single site but was repeated at several sites at different times. In countries where a system of herdbooks is not in operation, such as parts of India and Africa, one finds types of animals associated with different tribes and regions. They have evolved partly by natural selection to suit local environmental conditions and partly by artificial selection. Many of these local types, or land races, have well marked fancy points such as colour and shape of horns but they also differ in productivity. Domestication has produced a considerable range of animal types within each species. The development of the main cattle types is given in Fig. 8.1.

It is unfortunate that for most domesticated farm animals the wild populations from which they are descended have become extinct. Once the domesticated animals were found to be better than their wild relatives, man hunted and destroyed the wild populations so that they could not cross-breed with and spoil his improved stock.

Feral animals

When an improved type of animal which has been reared and selected under good nutritional conditions is allowed to go 'wild' and again is subject to natural selection, with conditions of variable and often low planes of nutrition, its appearance tends to revert to that of its original wild ancestor. In New Zealand there are no native wild pigs, but when Captain Cook went there in 1772 he took some English domestic pigs and let them go wild. These 'reverted' pigs are today very similar to the wild boar in both conformation and composition; they are short in the body and long on the legs compared with the improved type. Moreover, the limb bones are thin and long as compared with those of the improved type (Fig. 8.2) just as the bones of wild sheep are thin as compared with those of sheep that have been improved for meat qualities (see Fig. 5.24).

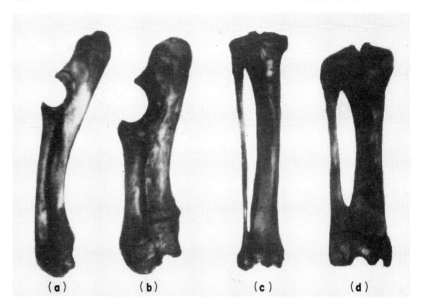

Fig. 8.2   The radius-ulna (**a**) and tibia-fibula (**c**) of a 'reverted' wild New Zealand pig compared with the radius-ulna (**b**) and tibia-fibula (**d**) of an improved pig of the same carcase weight, all shown to the same scale.

Breeds and conservation

During the eighteenth century and from among the variable land races, the early agricultural improvers selected animals which from their appearance and from simple measures of performances suited the production systems and the market requirements of the time. In order to distinguish the selected groups the breeders standardized the external appearance of the animals. This process of selection led to the isolation of separate strains, many of which had different but desirable performance traits. Breeders with similar objectives and with similar looking stock then grouped together to form a breeding cooperative or breed society for the organization of pedigree registrations, to ensure that the separate strains were maintained in reproductive isolation. Most of today's breeds were established during the late eighteenth and nineteenth centuries. Since then there has been considerable selection within and between breeds, and breed substitution has been commonplace in many parts of the world. The widespread replacement of local breeds by the Friesian and Holstein for milk production is a good example.

Just as the wild ancestors of modern breeds, and many of the land races, of farm animals have become extinct through man's activities, so many breeds of livestock are being lost. Those breeds whose performance is not as good as other breeds at meeting contemporary economic needs become less popular, their numbers decline and very often they become extinct. Concern is now being expressed at this loss of valuable variation. The economic needs of the years ahead are not known and a breed which is unpopular today may carry characters which will be most valuable in the future. To safeguard the maintenance of variation in animals, conservation is being carried out for farm breeds of livestock just as it is for the wild species. Surgical ovum transfer has been used for breed substitution; the technique of deep freezing of semen and embryos (p. 72) offers a relatively cheap method for conservation of presently uneconomic breeds.

Livestock improvement

Let us now consider the origin of the commercial characters of domestic animals. Under natural conditions wild animals are at the mercy of the environment, and in most places have abundance of food at one time of the year and little at another; consequently there is a regular seasonal variation in body weight. Very much the same conditions exist in the domestic animals of the less developed areas of the world and under some range conditions today (Fig. 8.3). The limitation of food intake can inhibit the development of body conformation in cattle (Fig. 8.4) giving them almost the appearance of a different breed.

With the development of agriculture, however, conservation of food has made it possible to keep animals on a high plane of nutrition all the year round. Thus, by modifying some of the environmental conditions, it has been possible to breed and select for animals which reach puberty and lay down fat at younger ages, which it would have been impossible to do under the natural conditions of intermittent food supplies.

With the increase of fodder and cereal supplies has come the opportunity and desirability of breeding types of animals which will produce more meat, milk, eggs and wool per animal or per unit of feed. A good nutritional environment is, in most cases, highly desirable for the selection of improved animals (but see pp. 226-8). The improved types of animals do not digest their feed more efficiently than the unimproved. The difference between them is mainly due to the way in which the nutrients are utilized by the body after absorption. For example, in order to make 45 kg liveweight gain the Poland-China breed of pig required only 189 kg of concentrate feed, whereas the first cross with the wild boar required 218 kg, and the backcross to the wild boar required 291 kg (Culberson and Evvard, 1926). An animal selected for meat or milk production therefore uses feedstuffs much more efficiently, in the sense that a higher proportion of the feed intake is utilized for production and a lower proportion is required for body maintenance. This is well illustrated from the facts shown in Fig. 8.5. The distribution of the nutrient intake to the different parts of the body such as the skin, bones, muscles, fat and udder is determined by the nutrient intake level (see p. 164) on the one hand and by the animal's genetic constitution on the other. Thus, if as a result of selection animals secrete more pituitary lactogenic hormone, or have mammary tissue which responds more sensitively to that hormone than their ancestors, the udder might be expected to be more developed and to utilize a higher proportion

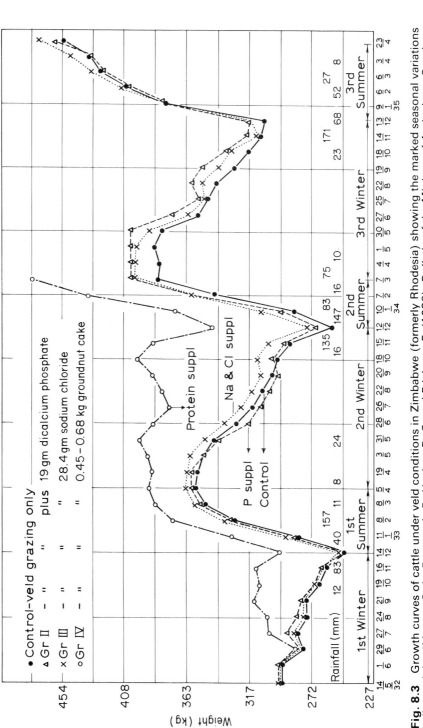

Fig. 8.3 Growth curves of cattle under veld conditions in Zimbabwe (formerly Rhodesia) showing the marked seasonal variations in live weight. (Murray, C. A., Romyn, A. E., Haylett, D. G. and Erichsen, F. (1936). *Bulletin of the Ministry of Agriculture, Southern*

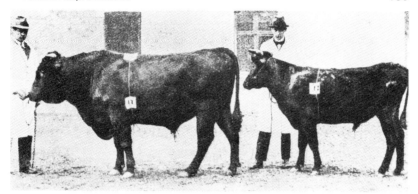

Fig. 8.4 The effect of the plane of nutrition during rearing on the size, conformation and male characters of Red Danish bulls by the same sire. (*Left*) Well reared: 3998 feed units fed; 2 years 45 days old; 707 kg live weight. (*Right*) Badly reared: 1387 feed units fed; 2 years 44 days old; 318 kg live weight. (Frederiksen, L. (1929). *Beretning Nordisk Lanbruksteknisk Kongress*, **4**, 67.)

of the nutritional intake. The control by the endocrine system of nutrient use for growth is explained in Chapter 2.

Breeding animals for commercial purposes such as meat, milk and egg production is quite different from the creation of 'fancy' breeds. With the latter, the breeder is only concerned with aesthetic considerations. He takes some new mutation and makes a breed of it, or includes the character in one or other of the existing breeds. By contrast, the breeder of horses for

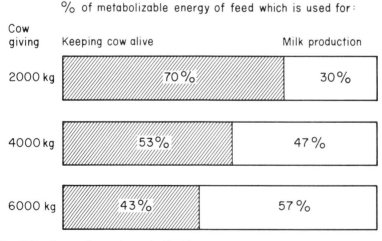

Fig. 8.5 Proportions of metabolizable energy used for maintenance and for milk production by cows with different annual yields. The figures are calculated for Friesian cows weighing 590 kg, giving milk with 3.7% fat and 8.6% solids-not-fat.

speed or draught, cattle for milk, or sheep for wool has a definite objective and directs the improvement of his animals. What this directed evolution means can be understood from the advice given to the prospective Merino flock master who 'must visualize the type he wishes to attain and the wool it is to grow. That "dream sheep" must be kept before his mind at all times'. As an example of what has been achieved by this method in Australia, the average fleece weight of adult sheep in New South Wales during 1881–5 was 2.3 kg while during 1956–9 it was 4.4 kg (Turner, 1962). However, this improvement is not due entirely to selection because during these years management also improved.

Another example of directional breeding is the Danish bacon pig. The present requirements for a good bacon pig are that it should have good feed conversion, length, thin backfat, light shoulders and large hams.

Table 8.1 Transmission of economic characters in crosses between the improved Danish Landrace pig and unimproved wild boar. (Clausen, H. (1953). *The Improvement of Pigs:* The George Scott Robertson Memorial Lecture, Queen's University, Belfast.)

Percentage of blood		Days to reach live-weight of:		kg barley required per 1 kg live-weight gain	Length of small intestine (cm)	Body length* (cm)	Rib number	Thickness of back-fat (cm)	Points out of 15 for	
Wild boar	Land-race	kg 20	kg 90						Shoulder	Ham
75	25	142	388	5.79	1381	83.7	14.4	3.86	5.0	3.9
50	50	107	256	4.44	1650	86.1	15.0	4.48	8.7	5.6
25	75	85	217	3.74	1755	89.4	15.8	3.94	10.9	9.0
0	100	76	180	3.06	2132	93.4	16.0	3.42	12.7	12.5

*From symphysis pubis to axis

These were first specified by Danish producers and they set out to select for these characters under suitable and controlled environmental conditions in progeny-testing stations, with considerable success. As the body composition of the pigs changed between 1924 and 1960, the feed conversion improved from 3.57 to 2.95. Again the improvements were not wholly due to selection because the management and environment in the testing stations were improved.

When an improved pig, such as the Danish Landrace, is crossed with the wild boar, from which it was originally derived, and the offspring are mated back in each direction, there is a gradual transition from good to bad qualities in animals of different proportions of the genes of the two parent breeds (Table 8.1). No qualities are completely dominant or recessive as they would be if they had been produced by sudden large mutations.

It is only in recent years that methods have been developed for measuring the amount of improvement achieved by selection separately from the improvement resulting from changes in management and the environment generally. For example, in the United Kingdom egg production per bird per year has increased from 120 in 1945 to well over 230 in 1975, representing an increase of over 3 eggs per bird per year. About half this improvement has been gained by changing the genetic merit of the birds and the other half from improvements in nutrition, housing, health and management. The change which directional breeding achieves is small each year but it is cumulative and over decades represents a considerable change in production.

References

CULBERSON, C. C. and EVVARD, J. M. (1926). Costly influences of an inferior sire. *American Herdsman.*

ISAAC, E. (1970). *The Geography of Domestication.* Prentice Hall Inc., Englewood Cliffs, New Jersey.

TURNER, HELEN NEWTON (1962). Production per head. In *The Simple Fleece.* (A. Barnard, Ed.) Melbourne University Press.

ZEUNER, F. E. (1963). *A History of Domesticated Animals.* Hutchinson, London.

Further reading

BOWMAN, J. C. (1977). *Animals for Man.* Studies in Biology, no. 78. Edward Arnold, London.

DARWIN, C. (1875). *The Variation of Animals and Plants under Domestication*, 2nd edition. John Murray, London.

FRIEND, J. and BISHOP, D. (1978). *Cattle of the World in Colour.* Blandford Press, Poole, Dorset.

9 Mendelian applications

The mechanism of inheritance

The observed or *phenotypic* characters in animals are the result of the genetic make-up or *genotype* of the animal and the environment in which it develops, lives and produces. Characters are passed on from one generation to another by many genes, together forming the genotype, which are on the chromosomes within the cell nucleus. The chromosomes are formed by very long molecules of deoxyribonucleic acid (DNA) and each gene is a segment of one of these DNA molecules. Variations in the composition of the sections of DNA molecules are the gene differences which determine the development of the animal's characters throughout its life.

In the body cells of an animal, each of the chromosomes is duplicated (see Fig. 9.1). The number of chromosome pairs is the same for almost all individuals of a species but is different for different species (Table 9.1). When the sex cells of an animal are formed, a reduction division, or *meiosis*, occurs and only one chromosome of each pair goes into the egg or the spermatozoon, so that only half the number of genes that are in the body cells of the animal go into the sex cells (Fig. 9.2). When fertilization occurs the corresponding members of each chromosome pair, one from each parent, come together, and so the genes are duplicated in the body cells of the developing embryo. It is in this way that the characters from the two parents are transmitted to their offspring.

Occasionally during the formation of the sex cells the division of the chromosome pairs is not perfect. Also, during the division of the body cells, the replication of chromosome pairs is not always perfect. The results of these imperfections are cells with abnormal numbers of chromosomes for the species, sometimes too many and sometimes too few (see Fig. 9.3).

Table 9.1 Normal chromosome number in farm animals; the numbers given include two sex chromosomes.

Horse	64	Goat	60
Donkey	62	Pig	38
Cattle	60	Chicken	78
Swamp buffalo	48	Mink	30
River buffalo	50	Dog	78
Sheep	54		

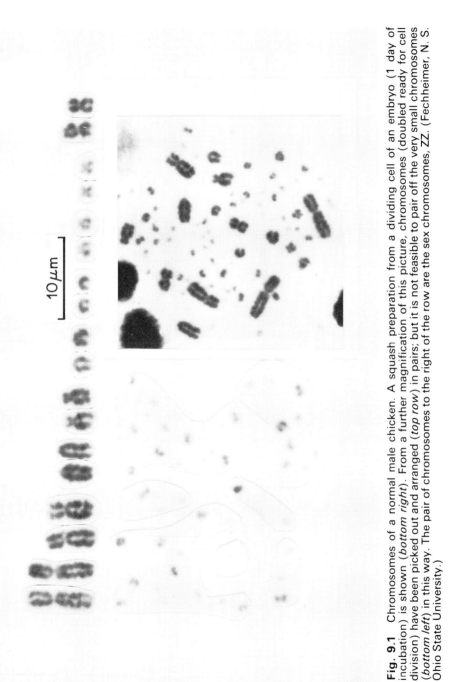

Fig. 9.1 Chromosomes of a normal male chicken. A squash preparation from a dividing cell of an embryo (1 day of incubation) is shown (*bottom right*). From a further magnification of this picture, chromosomes (doubled ready for cell division) have been picked out and arranged (*top row*) in pairs; but it is not feasible to pair off the very small chromosomes (*bottom left*) in this way. The pair of chromosomes to the right of the row are the sex chromosomes, ZZ. (Fechheimer, N. S. Ohio State University.)

10 μm

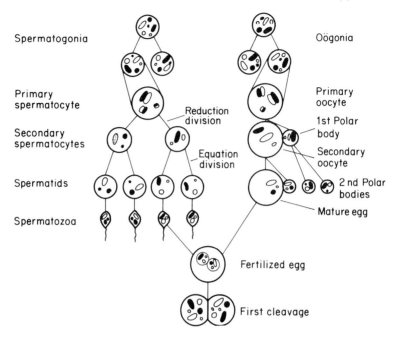

Fig. 9.2 The reduction division in the formation of sperms and eggs, showing how only one of each pair of chromosomes in the body cells of an animal pass to the germ cells. (Sinnott, E. W., Dunn, L. C. and Dobzhansky, T. (1958). *Principles of Genetics.* McGraw Hill, New York and Maidenhead.)

Cells with abnormal chromosomes can survive but the individuals composed of such cells may be defective, so that they are more prone to disease, they may not grow as quickly, or their productivity or reproduction may be poorer than that of normal individuals. Many chromosome abnormalities which arise during the formation of sex cells lead to the very early death of the embryos which develop from such cells. Thus many chromosome abnormalities are eliminated before they can appear in adults.

The special pair of chromosomes which determine sex are called the sex chromosomes. In mammals, the female has a pair of similar chromosomes called the X chromosomes and is termed the homogametic sex, while the male has one X chromosome and another quite different one called a Y chromosome, and is termed the heterogametic sex (see Fig. 9.1). It is clear that half the offspring of a female (XX) and a male (XY) will be XX, and therefore female, and half will be XY and therefore male. This accounts for the 50:50 sex ratio. In birds the sex chromosomes are designated by Z and W, the female is the heterogametic sex (ZW) and the male is the homogametic sex (ZZ).

The genes are located in definite places (*loci*), and chromosome maps (Fig. 9.4) may be made to show the distribution along the chromosome of the genes for particular characters. Characters which are controlled by genes on the same chromosome tend to be inherited together—they are

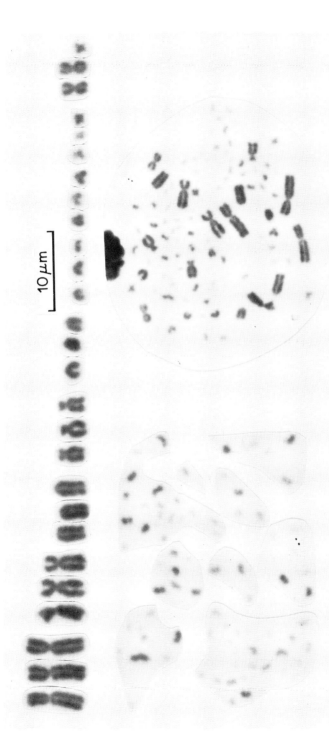

Fig. 9.3 Chromosome preparation from a chick embryo—preparation as for Fig. 9.1. This embryo is abnormal; it has three of every chromosome, and is called a trisomic. Its sex chromosomes (*top right*) are ZZW—two male and one female. (Fechheimer, N. S. Ohio State University).

10 μm

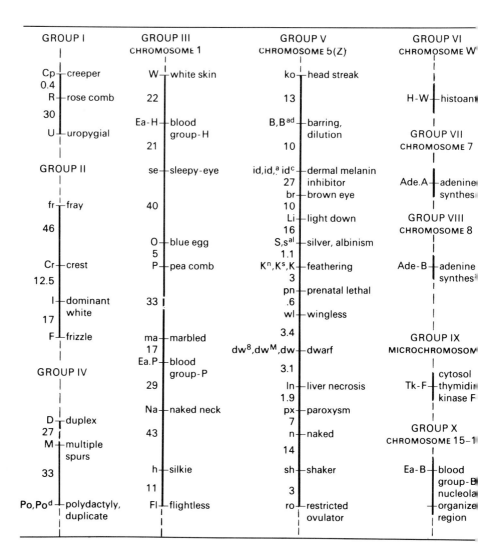

Fig. 9.4 Chromosome linkage map for the chicken, with 16 sex-linked loci and 24 autosomal loci in 10 linkage groups. Dominant mutants are indicated by initial capitals; numbers indicate relative distance between loci. (Somes, R. G. Jr., (1978). *Journal of Heredity*, **69**, 401.)

said to be *linked*. However, at meiosis (Fig. 9.2) this linkage may be broken by a process of breaking and recombining of chromosomes, which involves *crossing over* between loci on sister chromosomes. This means that it is remarkably difficult to discover which genes lie on the same chromosome. The linkage groups shown in Fig. 9.4 have been found only by long and laborious breeding tests. In the larger farm animals few linkages have so far been found, and these involve blood group genes and genes concerned with body mechanisms for disease resistance. Other known genes appear

to be transmitted independently—either they are on different chromosomes or they lie far apart on the same chromosome.

In an animal, *homozygous* for a particular character, both the controlling
genes at the same position (locus) on each of a pair of chromosomes are
exactly similar. When, however, the DNA constituting the gene at that
locus undergoes a change of composition a *mutation* occurs, and when the
sex cells of such an animal pair with the sex cells of a normal animal the

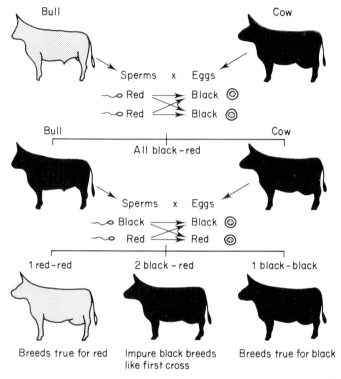

Fig. 9.5   Diagram to show how a pair of characters such as red and black
colour in cattle are inherited; red is recessive to black.

paired genes (*alleles*) of the embryo are now different, alternative, forms of
the gene, the normal and the mutant. Such *heterozygous* animals will not
breed true, for at the next reduction division half the sex cells will have the
normal gene and half will have the mutant form. If two such animals are
mated together (Fig. 9.5) the chances are that two normals will unite once,
two mutants once, and one normal and one mutant twice—giving an
offspring ratio of one homozygous normal, two heterozygotes and one
homozygous mutant.

For the most part these mutations are hidden (or *recessive*) in inheritance; that is, the heterozygous animal, which carries the mutant gene, is, in
appearance, like the homozygous (or *dominant*) form. New mutants, which

arise spontaneously with a very low incidence, are usually inferior to the normal alternative gene, and are eliminated from a population of animals by natural or artificial selection (see p. 218). Some of the mutations which appeared in the past have been used by breeders as a form of trademark to establish breeds.

Sex linkage

It is thought that there are very few genes on the Y chromosome in mammals (or the W chromosome in birds). Therefore genes which are located on the X chromosome (Z in birds) will be linked or associated with sex and will be transmitted to the next generation in combination with sex (Fig. 9.6). The fewer chromosomes the species has, and the larger is the X (or Z) chromosome, the more characters there will be associated with sex and inherited in a sex-linked way.

The practical application of sex-linkage occurs chiefly in poultry, in which it is a great convenience to be able to determine sex easily and early in life, so that pullets only can be reared when the chicks are wanted for egg production and the cockerels can be destroyed at hatching. The males of egg production strains of poultry are no use for broiler meat production. When the chicks are from strains to be used for broiler meat production the sexes can be identified and reared separately to improve growth rate and feed conversion.

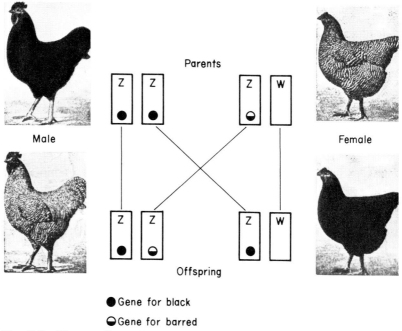

● Gene for black

◐ Gene for barred

Fig. 9.6   Diagram illustrating sex-linkage in poultry. When a black cock is crossed with barred hens, all the cocks come barred and all the hens come black.

The paired characters used for sex determination in poultry are early feathering as compared with late feathering, the down colours of silver as compared with gold, and whole coloured (dark heads) as compared with barred (light heads). When this sex-linkage in poultry was discovered, crosses between breeds were used for this purpose; such, for example, as between Rhode Island Reds (gold) and Light Sussex (silver). It was then discovered in crosses between two barred types (Plymouth Rock and Campine) that sex-linkage could be obtained within a breed breeding true for colour; such a breed (Cambar) was synthesized. More recently still it

Fig. 9.7 Sex-linked male (*left*) and female (*right*) chicks. (Pease, M. S. (1952). *Sex Linkage in Poultry Breeding.* Ministry of Agriculture Bulletin, No. 38.)

has been found that, provided the down colours of a bird are uniform, it is possible to make such a barred variety of any breed (Legbar, etc.) which will exhibit sex-linkage (Fig. 9.7).

Auto-sex-linkage has not fulfilled its early promise. Birds must be bred primarily for production, so breeders cannot afford to restrict themselves to the barred breeds. Sexing is now done mostly by examination of external genitalia of the newly hatched chicks.

It seems unlikely that any commercial quality is wholly sex-linked, for commercial qualities are not simple mutant characters and do not depend on a single gene but on a large number of genes; it is unlikely that these genes are all situated on the sex chromosome.

Recombination of characters

The method of recombination of characters has been used to a great extent in plant breeding; its application in animal breeding, however, has so far been very limited, not only because of the large numbers of individuals which must be eliminated in the process (and hence the cost), but also because of the time it takes to carry out. The following gives an example of recombination: we have a black-faced polled breed and a white-faced horned breed of sheep and wish to make a recombination, black horned or white polled, which will breed true. The first cross (F1 generation) gives speckled-faced (intermediate) horned (dominant) animals; if these first

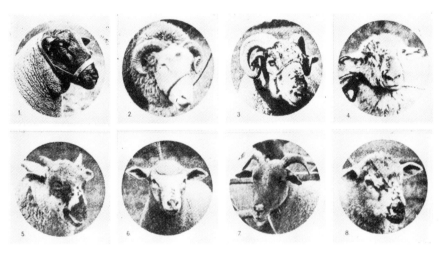

Fig. 9.8 Inheritance of horns and face colour in sheep. In this example horn development is male hormone dependent, so when a polled black faced breed **(1)** is crossed with a horned white faced breed **(2)** speckled faced horned males **(3)** and polled females **(4)** are produced. When the latter two are bred together, a range of types, **(5)** to **(8)**, including white faced polled **(6)** and black faced horned **(7)** animals are produced. (Wood, T. B. (1901). *Journal of Agricultural Science*, **1**, 364.)

crosses are now mated together to obtain the F2 generation, this will contain a recombination of characters from which the combinations required can be picked (Fig. 9.8).

This method of recombination is comparatively easy when both (as in the case in sheep quoted above) or one of the characters we wish to combine is of the simple mutant or Mendelian segregating type. Suppose, for example, it is desirable to make a polled Ayrshire having all the characters of the Ayrshire but being polled instead of horned. The milk characters of the Ayrshire are not of the simple Mendelian segregating type but show a blending form of inheritance (see p. 221). The horned and polled condition in cattle is, however, simple—polled being dominant to horned. If we mate a Red Poll bull to an Ayrshire cow we obtain polled

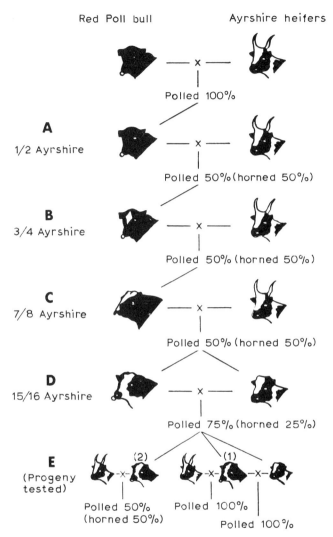

Fig. 9.9 Diagram showing how a character such as polled can be transferred from one breed to another. The polled character is dominant and so is passed on to 50 per cent of the offspring in each generation after the first. As most breed characters are inherited in an intermediate way, by picking out a polled bull in each generation and grading to pure Ayrshire heifers it is possible to produce a pure Polled Ayrshire in generation **D**. If these animals are bred together and bulls in generation **E** are progeny tested, on horned heifers, those bulls **(1)** that breed 100 per cent polled calves can be used to form a herd of pure-breeding Polled Ayrshires.

calves which will be heterozygous for the polled condition, but intermediate in body form. If we now mate a bull calf of this generation back to pure Ayrshire cows half his offspring will be polled, and all three-quarters Ayrshire in body type. We again pick out a polled bull calf from this generation and mate him to the pure Ayrshire as before, and the same thing happens again, except that the offspring will now be seven-eighths Ayrshire in type. This is repeated until all traces of other than Ayrshire body characters have disappeared. At this stage the polled animals should be bred together and in this generation there will appear pure polled Ayrshires in the ratio of 1 in 4. To find out which are pure for the polled character the bulls should be mated with horned cows and those which fail to produce any horned offspring (progeny-tested bulls) should be kept for future breeding. If such progeny-tested bulls are used in successive generations the horned character will be bred out. This is illustrated in Fig. 9.9.

Some indication of whether a bull is likely to be, and breed, pure for the polled character can be obtained from an examination of his skull. If he has small scurs or bony lumps under the skin where the horns should be (Fig. 9.10) he is unlikely to breed true for the polled character. That is, in males the polled character is not completely dominant and heterozygous males may have lumps, small scurs or even small loose horns, depending

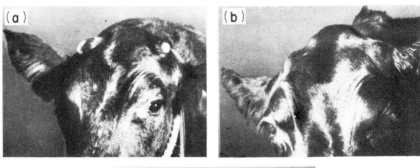

Fig. 9.10 Heads of steers showing **(a)** scurs, or loose horn growth, and **(b)** a small knob of bony growth under the skin; these are heterozygous forms of the polled condition. **(c)** Pure polled. (Hammond, J. (1950). *Endeavour*, **9**, April.)

on the breed. On the other hand, the polled character in most breeds is completely dominant in the female, although in a few breeds loose horns may occur. The heterozygous polled female therefore can only be detected by producing a horned or scurred male calf when mated with a pure horned or pure polled bull respectively. Polled strains have been, or are being, bred in many of the main horned beef and dairy breeds.

Blood groups

The importance of blood groups for transfusions of blood in humans is well known. In man, horses and pigs, but not in cattle or sheep, when certain blood groups are present in the parents the newborn may suffer from a haemolytic disease which causes severe anaemia. In cattle (and also in sheep, horses and pigs) the blood groups are much more complicated than in man. These blood groups are inherited—no gene (allele) can be present in a calf unless it is present in one or both parents. Thus the blood groups can be used to check pedigrees and investigations have shown that in addition to deliberate falsification of pedigree, error can occur through a cow which is pregnant to one bull having a subsequent heat at which she is served by another bull. Another use for blood groups is in tracing the affinity between different breeds in all species of livestock.

In addition to differences in the red blood cells there are inherited variants of the blood serum proteins, of haemoglobin, and of the proteins in milk. Several investigations have indicated that there may be connections between some of these variants and productivity. For instance, cows carrying blood of one specific group have, on average, a higher milk-fat than those lacking it. Again, some specific types of serum protein (transferrins) are associated with a higher milk yield. Some of these types can be assessed when the animal is comparatively young. It may be possible to select animals for productivity on the basis of the biochemical markers, so reducing testing costs and generation intervals. It is not known whether the productive advantage associated with these particular proteins is due to the genes responsible for them, or to other genes closely linked on the same chromosome.

Coat colour

The many varieties in coat colour in our domestic animals have all been derived by mutations from the normal wild-colour genes. In these, not only one but several mutations may occur, giving rise to a series (allelomorphic series) of different mutant genes, any of which may take their place as one of the pair at this locus of the chromosome. Thus, Adalsteinsson (1970) has shown that in Icelandic sheep 17 main colours of fleece, one white and 16 non-white, are distinguishable. There are three pigment types—tan, black and brown.

Tan (rust-red) pigment occurs only in the birthcoat of lambs, in kemp and outer coat fibres, and in fibres on hairy parts of the adult animal, whilst black or brown pigment may occur in all fibre types. White sheep either lack skin pigment completely or may have a little tan pigment. In non-white sheep, black or brown pigment may be present in addition to tan pigment. The inheritance of the tan pigment, which may occur in all the 17 main colour types, is not fully known. There are four colour patterns in non-white sheep; grey, badgerface, mouflon and grey mouflon, and are expressed whether the pigment is black or brown. In addition broken colour in non-white sheep can occur independently of pigment type and colour pattern. The 17 main colours are controlled by genes at three loci, A, B and S. At the A locus six alleles have been found. They are A_1, A_6, A_2, A_3, A_4 and A_5 which cause white, grey mouflon, grey, badgerface, mouflon and no pattern respectively. In this series inhibition of pigment is dominant to pigment production. At the B locus, the gene B_1 for black pigment is dominant to gene B_2 for brown pigment; at the S locus, the gene S_1 for unbroken colour is dominant to its allele S_2 for broken colour.

It is interesting to note that the A_1 allele has, in addition to the effect on coat colour, an adverse effect on fertility. This phenomenon of a single gene having multiple effects is called *pleiotropy*.

There may be two different genes giving rise to the same colour in a species. For example, in pigs, in most cases white is dominant to black, as, for example, in the Lincoln-Berkshire cross, but the white colour of the Mangalitza pig is recessive to black. Similarly in sheep, white wool is usually dominant to black, and black sheep are likely to occur in many white-woolled breeds; but in the Welsh breed there is a black which is dominant to white. This is the gene in the Black Welsh Mountain breed.

The white colour of albinos, which have pink eyes, is genetically quite distinct from the white colour of animals which have pigmented eyes. Physiologically an albino is an animal which lacks the capacity to form pigment, but it may contain the genes for any colour in its chromosomes. Albinism is recessive in inheritance, so when albinos are mated with coloured animals they produce young coloured according to the colour genes which they have inherited from their parents.

The coat patterns of animals behave in much the same way. A certain pattern may be present in an animal and yet not show itself until the colour genes are added, very much as the exposed film does not show up the picture until the developer is added. For example, the gene for striped coat is present in the wild boar, but in white and in black pigs it never gets a chance to express itself. When, however, black and white breeds are crossed and a mixture of white and black hair (roan) is present, then the striped pattern (Fig. 9.11) appears in many animals.

In some breeds of animal, such as Himalayan rabbits and Siamese cats, the shade of colour which is produced by the gene depends on the skin temperature. The animals are born white and the general surface of the body where the skin temperature is high remains pale, but the extremities (nose, ears, feet and tail) become dark. Dark colours on the body can be produced experimentally by exposing areas to cold. Thus the effect the

Fig. 9.11   A Large Black × Mangalitza sow with her litter by a boar of the same cross. In this case the black colour is dominant to white. The striped pattern of the wild boar that makes its appearance in the white pigling is caused by roan hairs. (Constantinescu, G. K. (1934). *Annales de l'Institut Nationale Zootechnique de Roumanie*, **3**, 13.)

gene has on the character can be changed by altering the physiological conditions under which it acts. In those animals such as the mountain hare and ermine which change from coloured summer to white winter coats the mechanism is quite different. This is a photoperiodic effect; it is the change in the hours of daylight which induces the moulting of the dark coat in autumn, and of the white coat in spring (see p. 28).

Complete dominance of the normal over the mutation is not always found, and in many cases the heterozygous animal (containing both nor-

Fig. 9.12   Diagram illustrating the inheritance of the red, white and roan colours in Shorthorn cattle.

mal and mutant gene) takes on an intermediate form and so can be distinguished from the pure dominant. There is an example of this in the genes for red and white coat colour in Shorthorn cattle, in which the heterozygote is roan. Roans, having on the chromosome pair one gene for red and one for white, will never breed true; and if mated together will produce calves in the proportion of 1 red, 2 roans and 1 white (Fig. 9.12). If one wishes to breed roans, the most certain way is to mate red cows with a white bull, or vice versa, for then all the calves will be roan.

Heterozygous breeds

Before the mechanism of inheritance was fully understood frequent attempts were made to fix these intermediate heterozygous forms and make separate breeds of them. For example, in the Blue Albion breed of cattle the standard set up as the colour of the breed was like that of the roan Shorthorn, but with black in place of red. The consequence was, of course, that large numbers of black and of white animals, not eligible for registration, were produced, and no amount of selection will ever increase the percentage of blue roans. Another case that may be quoted is that of the Blue Andalusian fowl which is the heterozygous form of the black and the white colour genes, and so will never breed true. A third case of a breed being founded on an intermediate heterozygous form is that of the Dexter breed of cattle. In the Kerry breed, from which the Dexter was formed, a mutation arose which, in the homozygous form, caused inhibition of the growth in the length of the bones and produced a bulldog calf. Such bulldog calves are not viable; they die and are aborted usually at about the 7th month. In the heterozygous condition, however, the gene for the bulldog calf produces merely a shortening of the legs. These animals had such an attractive appearance (Fig. 9.13) that breeders selected them and tried to breed them pure, with the result, however, that they continue to

Fig. 9.13 Calves of the short-legged heterozygous type (*left*) and long-legged homozygous type in the Dexter breed. The short-legged type contains the gene for 'bulldog' in the heterozygous condition; the long-legged type does not contain this gene. (Wilson, J. (1909). *Scientific Proceedings of the Royal Dublin Society*, **12**, Jan.)

produce calves in the ratio of 1 long-legged, 2 short-legged and 1 bulldog. Bulldog calves can be avoided, however, by mating short-legged cows with a long-legged bull and vice versa, when calves will be produced in the ratio of 1 long-legged to 1 short-legged. This is an example of how a mutant character can, even if it is lethal in the homozygous condition, be carried on from generation to generation in the heterozygous condition. A somewhat similar bulldog mutation occurs in the Telemark breed in Norway.

Abnormalities

A large number of mutations have occurred, and are occurring, in our domestic animals, but practically all of the visible ones give rise to defective characters or fancy points. Some can greatly affect body size and conformation; for example (Fig. 9.14) in mice recessive genes are known which cause dwarfing and adiposity (there is also—see below—another gene producing obesity). At first sight some of them may appear to be of real

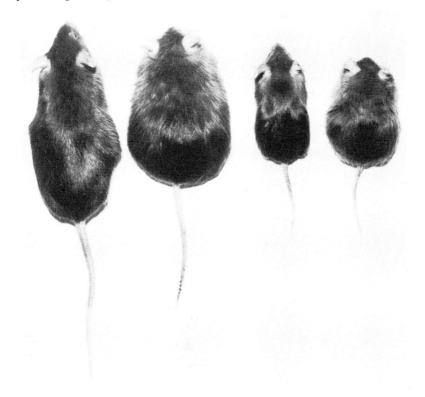

Fig. 9.14 The gene for 'adipose' and the gene for 'pituitary dwarf' have independent effects on the fatness and the skeletal size of the animal. From *left* to *right*: normal, adipose, dwarf and adipose dwarf mice. (Falconer, D. S. and Isaacson, J. H. (1959). *Journal of Heredity*, **50**, 290.)

economic value, for instance the double-muscled cattle (Fig. 9.15) have an increased proportion of muscle in the carcase, and correspondingly less fat and bone. Even this is of equivocal value, for the animals are unthrifty and infertile, (another instance of a pleiotropic gene) and so the character can only be carried on through the heterozygous form. However, the double-muscled bull can be cared for specially, and used for crossing on to normal cows to produce beef animals for slaughter only.

For the most part these mutations are the bane of the livestock breeder's life, for they give rise to degenerate and abnormal forms which detract

Fig. 9.15 A typical double-muscled Charolais bull aged 16 months. Note the prominent muscles with deep grooves between them which show up because of the thin skin and absence of fat. The rump is sloping and rounded and the tail is set high; the legs are short and the stance poor. (By courtesy of Dr B. Vissac, Laboratoire de Génétique Appliquée, Centre National de Recherches Zootechniques, Jouy-en-Josas, France.)

from the efficiency of the animal in producing milk, meat, wool etc. The literature on animal genetics contains many descriptions of such mutant characters. A few will be quoted to illustrate the range of characters which may be affected and some of the generalizations which can be made about them.

Many mutants give rise to abnormal development of the embryo leading to foetal death or atrophy. In poultry they often cause death in the shell during incubation (see Fig. 7.9). These are the so-called lethal factors. Lists of lethal and sub-lethal characters in domestic animals have been published. Such lethal factors may be associated with some other character and so prevent its appearance in a pure (i.e. homozygous) form. One gene for yellow colour in mice is lethal in the pure form, while in the heterozygous condition it causes extreme obesity. Likewise, dominant spotting in mice causes lethal anaemia when homozygous. The platinum gene in mink and silver fox is lethal in the homozygous condition. Dominant white is also probably lethal in horses—as in the Fredriksborg Stud of Danish horses.

White colour is often associated with other abnormalities. Hereditary deafness is common in blue-eyed white cats and in white dogs. White Shorthorn heifers often suffer from impaired development of uterus and vagina which renders them infertile. This is known as 'white heifer disease' and it also occurs in other white/roan coloured breeds such as the Belgian Blue.

Another interesting example of this pleiotropic effect of a single gene is the connection between hornlessness and intersexuality in goats. Homozygous polled females are intersexual (often wrongly called hermaphrodite), or look like sterile males. Therefore it appears impossible to obtain a purebreeding hornless breed of goats. Horns must be removed by chemical or electrical treatment of the horn buds. To produce normal polled goats (i.e. heterozygotes) one parent should be horned; half the offspring will be polled and half will be horned.

Mutations can be produced by treatment of the germ cell by radium or X-rays, which cause changes in the structure and the chemical composition of the genes. A very large proportion of such mutations, however, consist of gametic or early embryonic lethals, and thus lead to decreased fertility in the strain. Other mutations cause defects or abnormalities which make the animal less vigorous and less efficient commercially, and are of interest to the veterinarian in that they supply the explanation of cases he is often called on to treat. Examples of these are 'short-back' in cattle; atresia of the colon and sidebone in horses; earlessness and stiff joints in sheep; 'shakers' in goats; 'loco' in chicks; hydrocephalus, cleft palate, hernia, thickened legs, inturned nipples and imperforate anus in pigs; and cryptorchidism in goats, pigs, cattle and sheep.

Many of these abnormalities are due to two or more pairs of genes acting together rather than to simple recessives. Complications in inheritance may also arise because only some of the animals homozygous for the gene concerned may show its effect (*penetrance* is incomplete). In other cases penetrance may be complete and all homozygous animals may be affected. In many cases, also, *expressivity* is variable, i.e. the degree to which the abnormality shows itself in the individual varies greatly, although the gene which produces it is the same. This, for example, is so in a sex-linked lethal character (coloboma) which has been investigated in the chicken. In some embryos the abnormality may involve nearly all skeletal parts whereas in others it may be confined to a slight reduction of the upper beak (Fig. 9.16). The probable explanation of these variations in the appearance of the character is that the defective gene interferes with the rate of cell division and with the rate of embryo development. The later stages of normal embryo development are dependent on normal completion of the earlier stages. Disruption of the early stages gives rise to variable

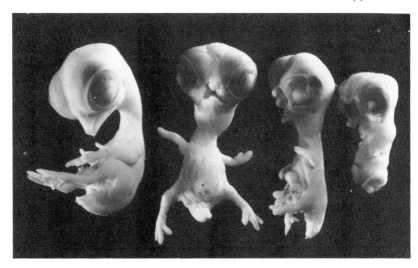

Fig. 9.16 *Left to right:* a normal chick embryo and three grades of expression of coloboma, mild, intermediate and severe. All embryos were sacrificed on the ninth day of incubation. (Abbott, U.K., Craig, R. M. and Bennett, E. B. (1970). *Journal of Heredity*, **61**, 95.)

abnormalities in the later stages. Thus, if the embryo is very slow in developing the abnormality is marked, whereas if the embryo is only slightly retarded the abnormality is slight.

A somewhat similar case of a recessive mutant gene giving rise to various degrees of an abnormality is that of the dropsical calf which occurs in the Ayrshire. Many of these have to be delivered by embryotomy and more are born dead than alive, although a few survive for a short time (Fig. 9.17).

Similar mutants occur in many different species; for example, the short dished face occurs not only in the bulldog, and in the Middle White pig (see Fig. 6.12), but has also appeared in cattle in the Argentine. 'Naked', or lack of hair, mutations have also occurred among normal animals of very different species—cattle and mice, for example. Other mutants such as dwarfing have been reported in the same species (cattle, Fig. 9.18) at different times and at different places.

Many of these mutations, when they have not been too harmful, have been seized upon by breeders who have used them as the basis for new fancy breeds. In this way, for example, many of the queer shapes found in the different fancy breeds of dogs have been formed. As has already been mentioned, the production of 'fancy' breeds of animals is quite a different matter from the improvement of commercial qualities or natural evolution. As the vast majority of mutations are recessive in inheritance the production of 'fancy' breeds is relatively simple, for animals which show the character will breed true to it.

As will be appreciated from what has been said, many of these mutations

Fig. 9.17 Live male dropsical calf (*left*) compared with a normal one. (Donald, H. P., Deas, D. W. and Wilson, A. L. (1952). *British Veterinary Journal*, **108**, 227.)

Fig. 9.18 Two miniature calves from grade Charolais cows and a Charolais calf of normal size and the same age. (Gregory, K. E. and Spahr, S. L. (1979). *Journal of Heredity*, **70**, 217.)

are not so hardy as the original normal forms (Fig. 9.19) and this has probably given rise to the idea that the 'improved breeds' are not so hardy and vigorous as the unimproved. This is true of many forms, such as the frizzle fowl, and the porcupine pigeon, but where the character affected is not such as to weaken the constitution of the animal, such, for example, as a colour change from grey to black hair, a fancy breed may be equally as hardy as the normal dominant type.

Fig. 9.19   Naked and normal littermate male mink at eight weeks of age. Note the defensive posture, smaller size, and wrinkled skin of the naked mink. Nakeds seldom survive more than ten weeks. (Shackleford, R. M. (1973). *Journal of Heredity*, **64**, 166.)

Eliminating unwanted characters

We will now consider how we can purge our commercial breeds of livestock from the unwanted mutations which lower their efficiency for production, causing the appearance of a number of individuals that are economically unprofitable.

Take, for example, the case of an inherited abnormality in pigs. As the unwanted characters are practically all recessive, if the young boars which are going to be the herd sires are mated with one or two sows which show the defective character, any boar which is heterozygous will give offspring half of which will show the character. Such boars can then be slaughtered. If, however, their offspring from affected sows are all normal then the boars can safely be used in the herd; and while such boars are used in the herd it is certain that the defect will never appear. It is better to start testing the males rather than the females on account of the larger number of offspring which they produce. Where the character is lethal the heterozygous female must be used; but in this case only 1 in 4 of the offspring of heterozygous males may be expected to show the character. This, however, is a counsel of perfection. A breeder would be better advised to get rid of all affected animals and their parents (which will be carriers) and to buy-in unrelated sires.

The prevention of the spread of unwanted characters throughout a breed is very important where males are used extensively, as in artificial insemination, in which one bull may sire sixty thousand calves a year. One way to guard against this is to mate the bull to be proved to 20 or more of his oldest daughters. If he carries any recessive defects, whether they be lethal, or of conformation or constitution, they will appear in the offspring. Sometimes if a defect becomes widespread in a breed it may be desirable to use special measures to reduce its frequency. In Galloway cattle a lethal defect known as tibial hemimelia syndrome (Fig. 9.20) has been found to affect between 1 and 2 per cent of purebred calves in Scotland. The defect is almost certainly caused by a recessive gene, and identification of heterozygous carriers is necessary to prevent them being used as parents of future purebred stock. The Galloway cattle society and the East of Scotland College of Agriculture have set up a special herd of cows which are all

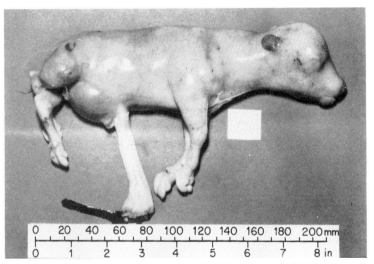

Fig. 9.20   A ninety day old cattle foetus with tibial hemimelia. (Pollock, D. L., Fitzsimmons, J., Deas, W. D. and Fraser, J. A. (1979). *Veterinary Record*, **104**, 258.)

known to have given birth to a calf with tibial hemimelia and therefore are heterozygotes. The purpose of the herd is to test-mate bulls which might be used widely in the breed. Each bull is mated to produce a minimum of ten calves, which will permit the detection of at least 94 per cent of heterozygous bulls. The defect is obvious in 90 day old foetuses, so in order to test more bulls in the special herd the cows are deliberately aborted at 90 days by prostaglandin treatment (Pollock *et al.*, 1979). This allows two test foetuses per cow per year, instead of less than one if the foetuses were allowed to continue for a full gestation period. In practice an artificial-insemination sire carrying a recessive defect usually shows up without special testing. There are enough carriers among the general female population to which he is mated; and one in four of their offspring will be abnormal if the bull is heterozygous for the same gene.

References

ADALSTEINSSON, S. (1970). Colour inheritance in Icelandic sheep and relation between colour, fertility and fertilization. *Journal of Agricultural Research in Iceland*, **2**, 3.

POLLOCK, D. L., FITZSIMMONS, J., DEAS, W. D. and FRASER, J. A. (1979). Pregnancy termination in the control of the tibial hemimalia syndrome in Galloway cattle. *Veterinary Record*, **104**, 258.

Further reading

BURNS, G. W. (1976). *The Science of Genetics, an Introduction to Heredity*, 3rd edition. Collier Macmillan, London.

DONE, J. T. (1976). Developmental disorders of the nervous system in animals. *Advances in Veterinary Science and Comparative Medicine*, **20**, 69.

HUTT, F. B. (1964). *Animal Genetics*. The Ronald Press Company, New York.

LEIPOLD, H. W., DENNIS, S. M. and HUSTON, K. (1972). Congenital defects of cattle: nature cause and effect. *Advances in Veterinary Science and Comparative Medicine*, **16**, 103.

PEASE, M. S. (1952). *Sex-linkage in Poultry Breeding*. 6th edition (1966), revised by C. M. Hann. *Bulletin of the Ministry of Agriculture, London*, No. 38.

RASMUSEN, B. A. (1975). Blood-group alleles of domesticated animals. In *Handbook of Genetics. Vol. 4, Vertebrates of Genetic Interest*, pp. 447–57. Plenum Press, New York and London.

SEARLE, A. G. (1968). *Comparative Genetics of Coat Colour in Mammals*. Academic Press, London.

10 Quantitative genetics and its application

Unit and multiple-factor characters

Up to this point those characters which are dependent on one gene only have been mainly considered. Unfortunately none of the commercial characters, such as body composition, growth rate, milk and butterfat yield, egg production, fleece weight, come under this heading. While unit factors (single genes) give rise to most of the defects and fancy points, all the commercial qualities investigated have been shown to be due to more than one gene; these are called multiple-factor characters.

Commercial qualities are in the main quantitative rather than qualitative and so cannot be measured as a fixed point, but must always be measured as a frequency-distribution curve about a mean. For example, although two breeds of cattle differing substantially in their mean butterfat percentages can be obtained, it is impossible to obtain a breed in which there is no variability in butterfat percentage. This defines it as a quantitative character which varies in amount not only with the genetic constitution but also with the physiological conditions and physical environment of the animal. When two breeds differing materially in their butterfat percentages are crossed together, the butterfat percentages of the resultant offspring fall on a distribution curve with a mean intermediate between that of the two parent breeds. When a second generation is raised by back crossing the first to the two parent breeds, the character is again intermediate, that is, there is no dominance or recessiveness, but the commercial character has a blending form of inheritance (Fig. 10.1). This is the usual way in which commercial characters like milk production are inherited. Physical characters, too, such as size, weight and body dimensions, are inherited in the same way.

Genotype and environment

The development of all these characters of economic importance is dependent on the environment in which the animal is kept and reared (Hammond, 1947). Changes in performance on farms are usually a combination of changes in environment and changes in the genetic quality of the animals. For example, the increase in the average milk yield for recorded

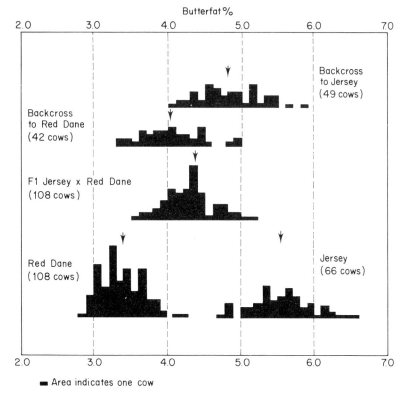

Fig. 10.1 Inheritance of butterfat percentage in crosses and back crosses between Red Danish and Jersey cattle. Arrow indicates average for each group. (Redrawn from Wriedt, C. (1930). *Heredity in Live Stock*. Macmillan, London.)

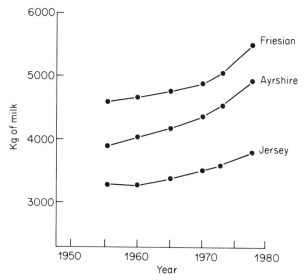

Fig. 10.2 Changes in herd average cow lactation yields; recorded herds in England and Wales, yields uncorrected for differences in milk quality. (Data from MMB Breeding and Production Division Reports.)

cows in England and Wales during the past thirty years (Fig. 10.2) is the result of improvements in nutrition, husbandry, disease control and the genotype of the cows. Some environmental factors such as age, dry period before calving and frequency of milking (2 or 3 times daily—see p. 87) affect the yield of milk given by a cow, and correction factors have been

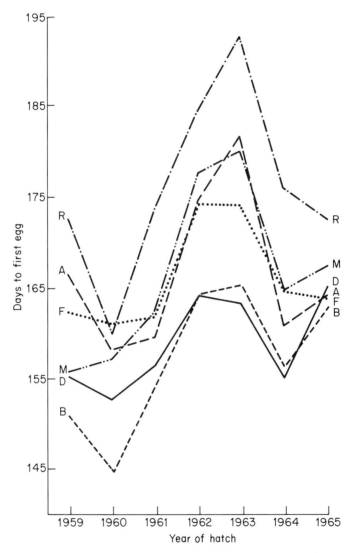

Fig. 10.3 Variation in age of sexual maturity (days to first egg) of six poultry strains, each bred to keep the same genetic composition from generation to generation. Differences between strains are genetic; fluctuations from year to year are due to changes in the environment, which affect strains similarly. (Bowman, J. C. and Powell, J. C. (1971). *British Poultry Science,* **12,** 511.)

calculated to correct yield for these factors, so as to obtain a better comparison between individuals differing in these respects.

Only by adopting special techniques or by setting up special tests is it possible to demonstrate the extent to which differences between individuals and breeds are the result of environment or genotype. For example, it is possible to breed populations of animals which remain of constant genetic quality over many generations. These 'control' populations can be used to measure differences in performance caused by changes in the environment of which the farmer may otherwise be unaware (Fig. 10.3).

The variation between animals shown in Fig. 10.1 refers to animals reared under the same conditions and fed in the same way. Even within a herd where each cow is given the same treatment or equal opportunity, where disease is absent and age of calving is identical, there still remain environmental differences between cows which often have more influence on the yield than do the genetic ones. It is these largely intangible effects within a herd to which the geneticist is referring when he speaks of 'environmental variation'. And it is the combination of the environmental effects and the effect of many genes, each with a small effect (multiple factors), which causes the variation in breed type shown in Fig. 10.1. The difference between the average of the two breeds and between the breed crosses, on the other hand, is entirely genetic.

Heritability

Some characters of an animal are inherited strongly while in others the intensity of inheritance is much lower because their development in the animal is more dependent on environmental conditions. The intensity of inheritance can be measured as the proportion of the superiority of the parents (over the herd or population mean) which is passed on to the next generation. This is called the 'heritability' of a character. The values of heritability are specific to the population on which they are measured. There may be substantial differences of heritability for the same character between populations with different breeding histories, or kept in different environments, as Table 10.1 shows.

The heritability of characters tends to be lower the more closely they are associated with reproduction. Thus in beef cattle the heritability of calving interval is about 5 per cent, of weaning weight is about 30 per cent and of carcass fatness about 50 per cent. This is a useful general guide to the level of heritability for all species.

It appears that the heritability of a character, and so the accuracy with which selection can be made, is greater when the environmental conditions are standardized, for then the differences in the expression of the character between individuals are more the result of heredity and less the result of environment.

The difference in the heritability of different characters, due to the degree to which they are dependent for their development on environmental conditions, is illustrated in identical twins in cattle, which, being derived

Table 10.1 Heritability of characters in beef cattle. There are up to 60 observed values per character, based on different breeds in different countries. The preferred values are those which are considered most accurate and at the median of the range. (Based on Preston, T. R. and Willis, M. B. (1970). *Intensive Beef Production*. Pergamon Press Ltd, Oxford.)

Character	Preferred value of heritability per cent	Range of observed values	Degree of heritability
Calving interval	5	2–20	low
Service period	5	3–13	low
Twinning	3	–	low
Gestation length	40	0–83	medium
Birth weight	38	0–100	medium
Pre-weaning growth rate (up to 6–8 months)	27	0–68	low–medium
Weaning weight	30	0–100	medium
Daily gain in feedlot	52	0–100	medium–high
Final weight	70	12–100	high
Feed intake	44	35–76	medium
Feed conversion	36	17–99	medium
Carcase fatness–fat thickness over longissimus dorsi	50	24–74	medium
Longissimus dorsi area	40	3–100	medium

from one egg, have the same genetic constitution. In such characters as colour, hair whorls and shape of horns, twins are very similar to each other (Fig. 10.4) because these characters are little affected by the environment. When, however, such twins are reared on very different planes of nutrition, the late-developing parts of the body can show considerable differences. Bonnier *et al.* (1948) found that the level of nutrition which was necessary

Fig. 10.4 Monozygous twin cattle, derived from the same egg and thus identical in genetic composition. (Courtesy of National Institute for Research in Dairying, Shinfield.)

to bring out the full genetical potential varied for different characters. The ceiling levels followed the growth gradients in that the head and forequarters were fully developed at a lower level of nutrition than that needed for full development of the hindquarters. Milk yield needed a lower level than complete body form.

Incidentally, identical twins can be used to measure the effect of nutrition, and of environment generally, independently of genetic influences. For example, using 120 pairs of twins and by placing one twin of each pair in a high-yielding dairy herd and the other in a low-yielding one, it has been demonstrated that the major part of the differences in milk and fat yield between the two groups of herds are due to the methods of feeding and management rather than to heredity (Brumby, 1961). Though identical twins have been used in the past to measure heritability, the values obtained in this way are usually substantially too large because of the large common maternal effects (see p. 230).

Selection

In any part of the body where variability exists, the selection and breeding together of the variant animals lead to changes in the character concerned. Thus, for example, while most sheep have only 2 nipples, some are occasionally found with 3 or even 4, and by selecting and breeding together such multi-nippled sheep it has been found possible to obtain ewes with 6 functional nipples. Similarly, while the pig has usually only 14 thoracic vertebrae and ribs, individuals occur in which the number is 15, 16 or even 17, and it is found that on the whole such animals have longer bodies. By breeding together animals showing this variation, strains with larger numbers of ribs may be produced. The higher the heritability of a character the more rapid the improvement expected from selection. A good example of what can be achieved has been shown by Hetzer and Harvey who, starting from a single pig population, selected one strain with more backfat and another with less backfat in each of two breeds (Fig. 10.5). After 10 generations of selection in Durocs the high strain had an average of over 5 cm of backfat thickness whereas the low strain had less than 3 cm. In Yorkshires, eight generations of selection resulted in a difference of over 1 cm backfat thickness.

Much has been learnt about selection methods by experiments on laboratory animals such as the fruit fly (*Drosophila*) and mice. The effects of selection for small and large sized mice in one such experiment can be seen in Fig. 10.6.

One of the first and most difficult aspects of selection is the choice of objectives. Not only is it necessary to select animals on the basis of characters which can be easily measured early in life and which are the main determinants of profit from the animal, but it also is necessary to have regard for the future market requirements perhaps 5–20 years hence. The improved animals produced by selection will be used in a future economic setting rather than the present one.

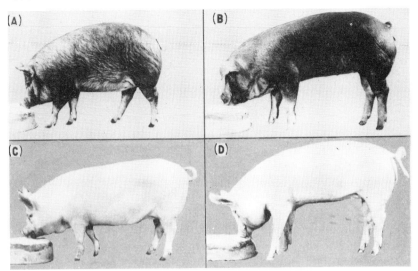

Fig. 10.5 Effect of selection for high and low back-fat thickness in Duroc and Yorkshire pigs. Thirteenth generation Durocs, **(A)** high and **(B)** low; eleventh generation Yorkshires, **(C)** high and **(D)** low. All at about 90 kg liveweight. (Hetzer, H. O. and Harvey, W. R., (1967). *Journal of Animal Science*, **26**, 1244; photograph by courtesy of B. Bereskin, United States Department of Agriculture.)

Selection is only possible under conditions in which a character is expressed. This applies for instance to milk production. Food intake is used for the maintenance of the body first, and only the surplus is used for production; consequently where food supplies are limited the genetic capabilities of the animal for milk production are not demonstrated and proper selection cannot be made. From this it has been concluded that when we wish to direct the improvement of an animal in any respect we should keep the population under selection in an environment which allows full expression of the character in question and rigorously select and breed from those individuals which show the character to the fullest extent. This is not necessarily so. For milk yield, for instance, bulls give a similar progeny test result whether tested in a poor environment (low-yielding herd) or a good one (high-yielding herd) (see Table 12.11). The advantage of testing in a good environment is that the heritability is usually higher and therefore fewer daughters are needed for an accurate progeny test.

Experiments on body size in mice have given equivocal results. Those by Falconer (1960) indicate that, for growth rate, it is better to select animals on a low plane of nutrition. Dalton (1967), on the other hand, got the same results whether he selected animals on a full diet or on one diluted with cellulose.

In view of this discrepancy between results, and of the differences between species and between characters, it is not safe to generalize either

Fig. 10.6   Effect of 32 generations of selection for high and low 6-week body weight in small populations of mice. *Left to right*; small, control and large mice at six weeks of age. (Falconer, D. S. (1973). *Genetical Research*, **22**, 291; photograph by courtesy of R. C. Roberts, Institute of Animal Genetics, Edinburgh.)

for or against selecting under optimal environmental conditions. However, there are records from most farm animals to show that the rank order of performance for breeds and families is not the same for different environments. Breeds have been developed to suit particular farming requirements. Therefore it is probably safest to select breeding animals under conditions which are as close as possible to those under which their progeny are to be exploited commercially (but see also Hammond, 1947).

Performance testing

In order to select the best parents for breeding, the animals available must be compared under standard conditions. Thus in order to compare animals from different herds, and therefore different environments, they must be brought together under the same conditions of feeding and management. Such an arrangement is called a 'performance test'. Its greatest use is for selecting males, and it can be used only for characters such as growth rate and feed-conversion efficiency which can be measured on a sire without slaughtering him. It is thus very important for evaluating beef bulls (see p. 289). With the development of ultrasonic methods of backfat measurement in pigs, performance testing of boars is tending to replace progeny testing (see p. 265).

Performance testing is measuring the characters in the sire himself, as contrasted to progeny testing which measures the characters in his off-spring. When a character is fairly strongly inherited (see p. 224) and where provision is made for a constant environment (including nutrition and management) the performance testing of males will give quicker, although less accurate, results than progeny testing; it will at any rate eliminate some males before progeny testing.

Progeny testing

The progeny test is the measurement of the genetic value of the animal in question by the production records of its offspring. Since a male produces during his lifetime many more offspring than a female, the progeny test is applied much more widely to males than to females. The progeny test is of the highest importance for those products, such as milk and eggs, which the male does not produce himself, but only transmits the capabilities of producing to his female offspring.

Progeny testing of sires is only justifiable if it increases the rate of genetic improvement per year compared with other methods of selection. It does so if the character under selection is *not* measurable in adult animals of both sexes, if its heritability is low, if the breeding unit is large, and if the increase in the generation interval, which is inevitable with progeny testing, is not disproportionately large. The first three of these criteria apply to milk yield in dairy cattle bred by artificial insemination; the increase in the generation interval is counterbalanced by the advantages of progeny test-ing. This is therefore the best method to use in selection for milk yield. In breeding for egg production in poultry, progeny testing increases the generation interval by 100 per cent; family selection therefore turns out to be the best method. Family selection is similar in practice to progeny testing except that the individuals chosen as parents of the next generation are the animals of the best families on test and not their parents as would be the case in progeny testing. The generation interval tends to be shorter for family selection than for progeny testing, but the number of superior animals which can be produced may be less. In pig breeding a further consideration is important. Here special testing facilities are necessary, which if used for sires rather than for progeny very much increases the number of sires which can be tested and therefore also increases the selection intensity among sires. With the development of methods for estimating the fatness of live pigs, the trend is to use more of the limited facilities of the test station for performance testing and less for progeny testing.

Maternal effect

One aspect of the environment which has a profound effect on some characters is the prenatal environment provided by the mother. Body size is particularly affected. When reciprocal crosses are made between the large Shire horse and the small Shetland pony (see Fig. 3.7), the size of the offspring in each case is considerably affected by the size of the dam. Similarly the mule with the horse as dam is a larger animal than the reciprocal cross, the hinny with the donkey as dam. This maternal effect on size is made use of in mule breeding, where the most desirable mules are bred by first crossing the large Percheron mare with the smaller but more active and spirited Thoroughbred stallion; the large active mare so produced is mated with a donkey stallion to introduce the character of heat resistance; thus large, active, heat-resisting mules are obtained. Maternal effects on size have also been shown in reciprocal South Devon × Dexter

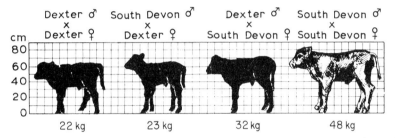

Fig. 10.7 Diagram showing average size differences in new-born calves of pure Dexter, pure South Devon and their reciprocal crosses. The size of the crossbred calf is limited in the large mother by heredity, whereas that in the small mother is limited by nutrition. (Joubert, D. M. and Hammond, J. (1958). *Journal of Agricultural Science*, **51**, 325.)

cattle crosses (Fig. 10.7). Maternal effect on size in rabbits has been strikingly demonstrated by O. Venge, who transplanted the fertilized eggs of a small breed of rabbit into a doe of a large breed and the eggs of a large breed into a doe of a small one (Fig. 10.8). Maternal effects are also shown in Border Leicester × Welsh sheep crosses and in transplanted ova (see Fig. 5.13). These maternal effects last for some considerable time after birth (see Fig. 5.14) in both sheep and cattle. In horses they persist up to adult life because length growth below the knee and hock is completed at birth. In other species where the proportional length of the lower limbs is not so well developed at birth there is scope, according to the state of its development at birth, for the growth to be made up by good feeding.

Adaptation to environment

In addition to selecting for productivity under ideal conditions, the breeder of livestock has also to breed for constitution, or the power to live and thrive under the conditions in which the animal is kept. Under natural

Fig. 10.8 Maternal effects on size in rabbits 6 weeks old. (*Above*) Young derived from a fertilized egg of a small breed which had been implanted into a doe of a large breed. (*Below*) Young derived from fertilized eggs of a large breed which had been implanted into a doe of a small breed. (Venge, O. (1950). *Acta Zoologica, Stockholm,* **31**, 1.)

conditions selection for this is by elimination of the unfit. This is one of the reasons for the occurrence throughout the world of so many local breeds of sheep, for the conditions under which sheep are kept come nearer to natural conditions than those for most other varieties of our livestock.

Let us now consider some examples of this. While European breeds of dairy cattle have a constitution eminently suitable for life in temperate climates, when they are transported to hot and humid tropical climates it is found that their tolerance of heat is limited. Their body temperature and rate of respiration go up (Fig. 10.9) and their appetite declines. As a result of this poor heat tolerance, much more time is spent lying in the shade rather than grazing (Fig. 10.10). Under these conditions of low food intake or utilization, growth is retarded, the age of puberty is delayed and milk yields decline; in fact in a few generations the breed degenerates (Fig. 10.11). In addition, tropical climates exert much of their adverse effect on production indirectly by lowering the quantity and quality of pasture and fodder crops available. Zebu cattle, however, which have developed in the tropics, have a constitution suited to their surroundings (good temperature regulation in hot humid climates and ability to thrive on low-quality

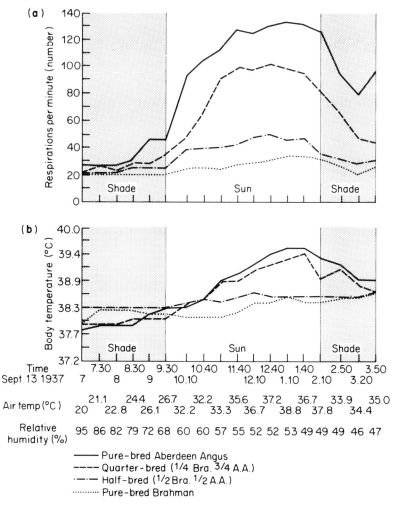

Fig. 10.9 Respiration rate **(a)** and body temperature **(b)** of cattle removed from the shade and exposed to strong sunlight on a summer day. Note the rise in the case of all except the purebred Brahman, which maintained almost a level rate throughout the period of observation. (Rhoad, A. O. (1936). *Journal of Agricultural Science*, **26**, 36.)

fodders) but are poorly developed in the milk-secreting tissues of the mammary glands. When crosses are made between European and Zebu cattle under humid tropical conditions the offspring grow better and yield more milk than either parent because from the one they receive udder development, while from the other they receive the constitution which enables the body to supply to the udder the nutrients required for milk production (Fig. 10.11). The severity of the climatic conditions will naturally determine the relative proportions of blood of each which are required to give the optimum results; lifetime yields are, of course, a better criterion than maximum lactation yields. The Indian Army Dairy Farms obtained

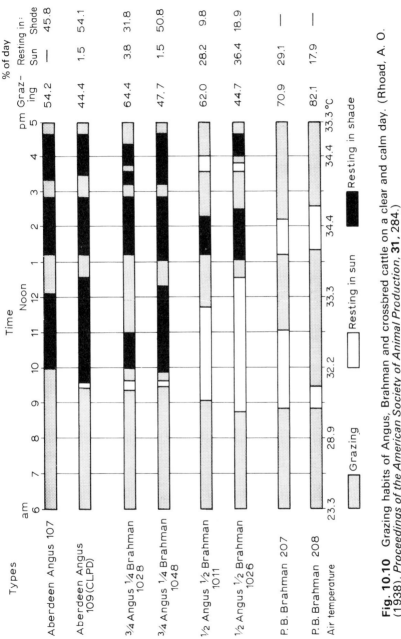

Fig. 10.10 Grazing habits of Angus, Brahman and crossbred cattle on a clear and calm day. (Rhoad, A. O. (1938). *Proceedings of the American Society of Animal Production,* **31**, 284.)

Table 10.2 Production of European × Indian Zebu cows at Military Dairy Farms during 1934–61. (From Amble, V. N. and Jain, J. P. (1967). *Journal of Dairy Science*, **50**, 1695.)

Grade of cow	First lactation		No. of female calves	Percentage mortality (birth to 1st calving)
	No.	Yield (kg)		
Sahiwal	118	1776	61	28
$\frac{1}{4}$ European	24	1573	37	32
$\frac{3}{8}$ European	96	2121	79	41
$\frac{1}{2}$ European	117	2561	24	4
$\frac{5}{8}$ European	72	2350	82	2
$\frac{3}{4}$ European	307	2335	194	18
$\frac{7}{8}$ European	218	2240	237	30
$\frac{15}{16}$ European	95	2109	154	36
$\frac{31}{32}$ European	10	1839	73	45

Fig. 10.11 Degeneration in a high-grade Jersey cow (*above*) in the tropics. Her daughter (*below*) by a Sahiwal Zebu bull is not only of better constitution, but has also given higher milk yields. (Hammond, J. (1932). *Report on Cattle-breeding in Jamaica and Trinidad*, Publication No. 58, Empire Marketing Board, London.)

such a striking improvement by making crosses of native cows with imported Ayrshire and Friesian bulls that they continued the introduction of European blood. There came a time, however, when the percentage of European blood became too high, and then production fell instead of increasing. By introducing into this 'degenerate' stock more Zebu blood (Table 10.2), and so a constitution suited to the tropics, production is increased (Fig. 10.11). In beef cattle, experiments along these lines have taken place in Queensland, Australia, and also in the southern states of the U.S.A. In both countries the first cross between Zebu and British breeds is strikingly superior to either pure breed. In Texas a new breed, the

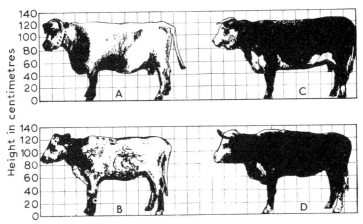

Fig. 10.12 Changes in type due to being grown under adverse high-temperature conditions. Shorthorn cow **(A)** under optimum conditions and **(B)** under high-temperature conditions. Hereford cow **(C)** under optimum conditions and **(D)** under high-temperature conditions. (Bonsma, J. C. (1940). *Bulletin of the Department of Agriculture and Forestry, University of South Africa*, No. 223.)

Santa Gertrudis, has been produced by combining the beef qualities of the Shorthorn (five-eighths) and the heat tolerance of the Zebu (three-eighths) (see Fig. 11.9).

Some characters in animals which may be an advantage in one climate may be a great disadvantage in another. For example, the thick winter coat in certain breeds of cattle holds a blanket of warm air round the skin and this, unless exposed to wind, insulates the body from cold. When such breeds are taken into tropical climates, however, many individuals fail to go into a short summer coat and so fail to regulate body temperature and do not thrive (Fig. 10.12). The reason why the moult of the winter coat of many British breeds fails to occur in the tropics is that it is controlled by photoperiod (Fig. 10.13). High temperatures also affect pregnant mothers in breeds that are not heat-tolerant, so that calves and lambs born after a very hot period are smaller than normal. It has been shown that one of the reasons why Zebu breeds are more heat-tolerant than British breeds is that

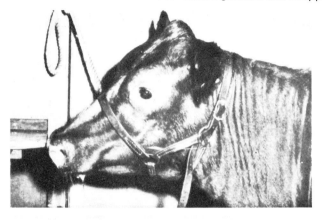

Fig. 10.13 Polled Shorthorns after 3 hours' exposure to a temperature of 40.6°C. (*Top*) Having had a period of increased light it has gone into a short summer coat and shows no distress. (*Bottom*) Having had a period of decreased light it has developed a thick woolly coat and shows heat stress by salivating. (Yeates, N. T. M. (1955). *Australian Journal of Agricultural Research*, **6**, 891.)

they sweat more. Selection for heat-tolerant characters in British breeds of cattle to suit them for the tropics is now taking place in Queensland. Heat tolerance is measured by the rise in body temperature when the air temperature is raised, or after exercise. Such tests show that, of European breeds so far tested, the Jersey is one of the best in this respect (see Chapter 4). The heat-regulating powers are not so well developed in the young animal as in the old one, and so selection for heat tolerance is best carried out in the calf stage.

Another example of the poorer heat-regulating powers of the young animal is seen in the pig, where owing to lack of coat covering the young animal suffers much from cold and draughts and so fails to grow properly (Fig. 10.14), the critical temperature being 15.5 °C for old pigs but 24 °C for young pigs. In pigs the subcutaneous layer of fat forms an insulator against heat loss. In newborn pigs there is little subcutaneous fat so that

Fig. 10.14 Pigs from similar litters at 11 weeks old. (*Above*) Reared in a cold sty. (*Below*) Reared in a warm hut. (Howie, J. W., Biggar, A. W., Thomson, W. and Cook, R. (1949). *Journal of Agricultural Science*, **39**, 110.)

deaths from cold are often considerable particularly where insufficient milk is available during the first few hours after birth.

The effects of climate on the animal's performance are very great and play an important part in the evolution and selection of domestic animals. The climate and environmental conditions of an area play a large part in the success of selection for any particular character. They also form one of the causes for the regionalization of different types of commercial livestock production.

Disease resistance

Among small animals many cases of differences in susceptibility to disease in different strains have been found. (This situation must be sharply distinguished from the case of constitutional diseases caused by mutations and hence behaving as unit factors in inheritance—see p. 213.) In the sphere of infectious or contagious diseases, certain strains of mice are highly resistant and others highly susceptible to leukaemia, mammary cancer, pseudo-rabies and mouse typhoid. In all such cases care has to be taken to distinguish between true genetic resistance and the immunity which is acquired by the offspring *in utero* or through the mammary secretions of the parent.

There is good evidence that susceptibility to some diseases in cattle is inherited, for example mastitis and Johne's disease, but the heritability is low. In our larger farm animals the process of obtaining resistant strains by selection in an environment of disease would take too long and would be far too expensive to warrant consideration. Furthermore, the reproduction rate in bacteria is high, and they can produce new virulent strains by mutation quicker than immune strains can be produced in domestic animals. African and Asian Zebu cattle have not developed immunity to prevailing diseases such as rinderpest and trypanosomiasis. They seem to

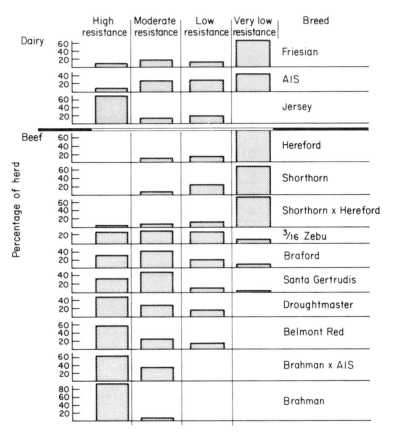

Fig. 10.15 Frequency distribution of resistance to cattle tick (i.e. extent of tick burden) in heifer herds of various breeds. The Brahman is an American humped breed probably derived from crosses of several Indian breeds of *Bos indicus*; Braford, Belmont Red, Droughtmaster and Santa Gertrudis are crosses of *Bos taurus* and *Bos indicus*. The AIS (Australian Illawara Shorthorn) may have a few *Bos indicus* genes. (Modified from Utech, K. B. W., Wharton, R. H. and Kerr, J. D. (1978). *Australian Journal of Agricultural Research*, **29**, 885.)

be less susceptible to tick infestation than European cattle (Fig. 10.15) and more tolerant of the tick-borne fevers, but they still need natural infection as calves to acquire immunity as adults. Some African humpless cattle such as the West African Dwarf Shorthorn and the N'Dama appear to be tolerant of trypanosomiasis.

A special and interesting case is afforded by cancer of the eye which occurs in white-faced European cattle (chiefly Herefords) in the tropics. Investigations in Texas have shown that it is strongly inherited and further-more that cancer of the eyelid (but not of the eyeball) occurs only on unpigmented eyelids. The disease can thus be reduced by using 'cherry-eye' Herefords.

References

BONNIER, G., HANSSON, A. and SKJERVOLD, H. (1948). Studies on monozygous cattle twins. IX. The interplay of heredity and environment on growth and yield. *Acta Agriculturae Suecana*, **3,** 1.

BRUMBY, P. J. (1961). The cause of differences in production between herds. *Animal Production*, **3,** 277.

DALTON, D. C. (1967). Selection for growth in mice on two diets. *Animal Production*, **9,** 425.

FALCONER, D. S. (1960). Selection of mice for growth on high and low planes of nutrition. *Genetical Research*, **1,** 91.

HAMMOND, J. (1947). Animal breeding in relation to environmental conditions. *Biological Reviews*, **22,** 195.

Further reading

BOWMAN, J. C. (1974). *An Introduction to Animal Breeding*. Studies in Biology, No. 46. Edward Arnold, London.

FALCONER, D. S. (1981) *An Introduction to Quantitative Genetics*, second edition. Longman, London and New York.

HAFEZ, E. S. E. (Ed.) (1975). *Adaptation of Domestic Animals*, third edition. Balliere Tindall, London.

WARWICK, E. J. and LEGATES, J. E. (1979). *Breeding and Improvement of Farm Animals*, seventh edition. McGraw Hill, New York.

YEATES, N. T. M., EDEY, T. N. and HILL, M. K. (1975). *Animal Science: Reproduction, Climate, Meat and Wool*. Pergamon Press, Rusticutters Bay, New South Wales.

11 Inbreeding and crossbreeding

Inbreeding

The mating together of close relatives, or inbreeding, not only makes more homozygous the germ-plasm and qualities of those animals which are inbred, but it also brings to light any recessive mutations which the strain of animals has in its germ-plasm.

The rapidity of the degree to which the germ-plasm is made homozygous and the uniformity of the animals obtained, depends on the closeness of the inbreeding; this comes in the following order—full brother to full sister or parent to offspring; half-brother to half-sister or grandparent to grand-child; nephew to aunt or niece to uncle; first cousins. The essential in inbreeding is an ancestor common to both parents. The farther back in the pedigree the common ancestor lies, the less intense is the inbreeding.

Linebreeding, although often used to mean 'mild inbreeding' is, correctly, the form of inbreeding based on a single common ancestor, for example a sire whose sons, grandsons and great-grandsons are used in succeeding generations. It may be successful, not because it is a form of inbreeding, but because of the exceptional merit of the animal on which the line-breeding is based. Its advantages are that a high degree of uniformity of type and production is attained in the herd. One form of linebreeding which has been successfully used with cattle is that of using home-bred rather than purchased bulls, the home-bred bulls being taken in turn from different cows which are known to have bred well. When after a time it is observed that the breeding has become too close, the first signs of this usually being a drop in milk yield and slow growth of the young animals, the best way to introduce new blood without disturbing type is to obtain a bull with 50 per cent outside blood. That is, to go to a good herd to which a good bull has been sold and buy a young bull sired by him out of an old high-producing cow which has already produced good milking daughters. To obtain new blood for beef herds which have become too inbred, and to prevent type being broken, the best way is to buy some good heifers and mate them with the herd bull. After noting how their offspring blend with the herd-type, a young herd bull can be bred from the heifers which breed nearest to the desired type.

By inbreeding, all recessive mutations lying hidden within the strain are brought to light. What these defects are will vary with each strain of animal inbred. For example, in two of four different strains of rabbits each inbred over some 20 generations few, if any, defective characters were seen, while in the other two strains a number of defective recessive mutations such as hypospadias, spina bifida, furlessness (cf. Fig. 9.19) and foetal atrophy appeared, but could be weeded out by selective breeding. These defects appear more frequently in inbreeding than in random mating because under inbreeding conditions the same allele is more likely to be present in both parents; defects are more likely to be recessive than dominant, so the same mutation has to be present in both parents before it can show itself.

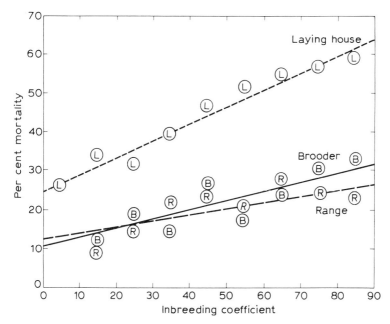

Fig. 11.1   Effects of inbreeding on viability of chickens. Unweighted linear regressions, from a series of 25 inbred lines of White Leghorns, of mortality in three stages of life (chicks in brooder, growing birds on range, and adults in laying house) on inbreeding. (Data of MacLaury, D. W. and Nordskog, A. W. (1956). *Poultry Science*, **35**, 582; from Lerner, I. M. (1958). *The Genetic Basis of Selection*. Wiley, New York; Chapman & Hall, London.)

Because of the many recessive genes with small negative effects on vigour, inbreeding invariably leads, in the long run, to inbreeding depression or degeneration (Fig. 11.1). Apart from specific defective mutations, the first signs of loss of commercial characters due to inbreeding appear in poor mothering qualities of the dam, loss of fertility, reduced milk production and slow growth of the young. Characters of less importance for survival, for example butterfat percentage, egg weight, and carcase characters, are less affected by inbreeding. King (1967), for example, found that while outbred Large White gilts produced on the average 10.2 young and weaned 8.3 each of 13.3 kg weaning weights, litters from 44 per cent inbred

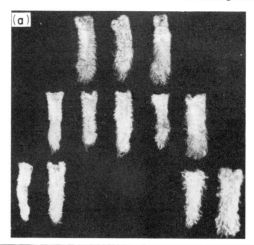

gilts averaged only 7.6 at birth and 5.6 weaned, with a weaning weight of 10.8 kg. For this reason, it is advisable, when exploiting hybrid vigour by crossing inbred lines, to use as female a cross between two inbred lines while using an inbred male. A male inbred for good qualities will breed better than he looks, whereas an outbred male will not breed as well as his appearance suggests. In the performance testing of males this has to be taken into account.

The first effect of inbreeding is to sieve out all the recessive genes that the animals contain so that these characters make their appearance usually in the first to third generation of inbreeding. Thus the first result of inbreeding is usually to cause increased variation in the offspring. If, however, proper selection of like types is made at this stage and inbreeding with the selected animals is continued, true breeding types will be obtained. For example, Fig. 11.2(a) shows on the top line the range of variability normally found in lambs' tails in a Romney flock, ranging from those in which the good wool spreads about half-way down the tail (left) to those in which the hair extends to seven-eighths up the length of the tail. On inbreeding the extent of the variability increases (see second line) due to the segregation of the recessive genes. Some of these recessive genes give rise to tails in which the good wool extends almost to the tip and other recessive genes give rise to tails which are hairy right up to the top of the tail. If selection is now made and a ram and ewes with good wool tails (bottom left) are mated together they will breed true for this character and the variability in the tails will be reduced. The same thing will happen if the lambs with hairy tails (bottom right) are bred together. This experiment demonstrates that inbreeding decreases the variation within inbred lines while increasing the variation between them.

The chances of obtaining the characters one requires on inbreeding are much increased if one inbreeds to progeny-tested sires or dams, that is, those which are known to have a large number of the desired genes. This

Fig. 11.2 How characters can be separated out and bred true by inbreeding and selection.
(a) *Top line:* the extent of the variation in wool and hair on lambs' tails found in an outbred flock of Romney Marsh sheep. *Middle line:* on inbreeding the extent of the variation is increased; in some (*left*) the wool extends almost to the tip of the tail, on others (*right*) the hair extends almost to the base of the tail. (*Bottom line*): when those with the woolly tails (*left*) are bred together they produce offspring with woolly tails (b) while those with the hairy tails when bred together also breed true to this character (c).
(b) Crop of lambs from the woolly-tail inbred line.
(c) Crop of lambs from the hairy-tail inbred line. The hairy tail is associated with coarse britch wool; the woolly tail with fine britch wool.

was, in fact, the method used by the pioneer breeders, as the following pedigree of the noted Shorthorn bull Comet bred by Collings shows:

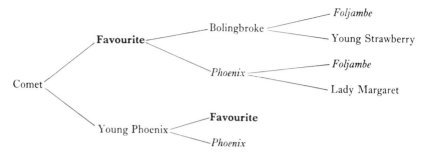

However, the inbreeding in the early history of the Shorthorn breed apparently also led to low fertility.

Studbooks, herdbooks and flockbooks

When beginning directive breeding for any particular object it is necessary first of all to find and then to concentrate the germ-plasm which contains the genes necessary for the development of the characters required. Studbooks, herdbooks and flockbooks have been established to segregate selected animals and breed them together, so ensuring that their germ-plasm is not diluted with the germ-plasm of unregistered stock. This concentration of the blood of strains showing the most desirable qualities involves restriction in the number of parents and hence inevitably leads to some inbreeding. Measurements of the coefficients of inbreeding show that with the passing of time the animals recorded in a studbook become more and more inbred (Fig. 11.3). Many of our herdbooks have now grown so large, however, that progress in concentrating the best germ-plasm is slow, and therefore some herdbook authorities are forming an 'inner circle' of animals with the best records—'advanced registers' within the herdbook—and it is from these animals that most of the sires used in the breed come.

Studies made on herdbooks show that in many breeds there is a stratification of herds. At the top are those of constructive breeders who are actively improving their cattle; they sell bulls to other pedigree breeders who multiply their stock and sell bulls to commercial producers. This is very evident in a breed such as the Friesian, where bulls have been imported. Recently, however, by the use of artificial insemination, it has been possible to shortcut this process by using bulls from constructive breeders direct on commercial cattle.

In interpreting the value of a pedigree from a herdbook, too much weight should not be placed on the performance of animals far back in the pedigree. The sire and dam each contribute half the germ-plasm of the animal, the grandparents one-quarter each, the great-grandparents one-eighth each (Fig. 11.4), and so on back in each successive generation. Figure 11.4 also illustrates how it is that, if the value of the sire and dam is known, more distant ancestors cannot give further information because their contribution to the individual's germ-plasm is included in that of the parents.

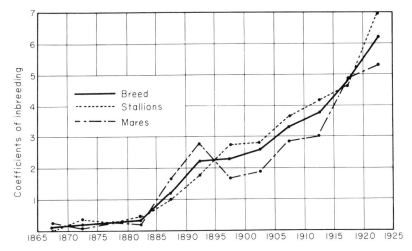

Fig. 11.3 Average percentage of inbreeding of horses entered in the Clydesdale Studbook from 1865 to 1925. (Calder, A. (1927). *Proceedings of the Royal Society, Edinburgh,* **47**, No. 8.)

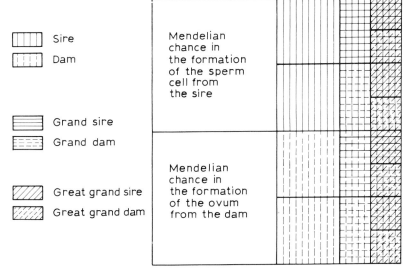

Fig. 11.4 Relative part played by the different ancestors in the genetic make-up of an animal as calculated from the Mendelian theory when random breeding is practised. (Lush, J. L. (1949). *Animal Breeding Plans.* 3rd edition. Iowa State College Press.)

Crossbreeding

Crossbreeding does exactly the opposite of inbreeding. In the first cross the dominant genes of the two breeds crossed mask the recessive genes. The first cross animal usually exhibits hybrid vigour because the recessive characters tend to be the less desirable ones. When, however, such first crosses are mated together, increased variability in appearance results because the recessive characters segregate again, so that animals may appear with the characters of both parent breeds. For example, in crosses between Aberdeen-Angus and Hereford cattle (Fig. 11.5) the black face of the Aberdeen-Angus and the horns of the Hereford behave in inheritance as do many of the bad characters of the animal and disappear in the first cross, which gives an animal with all the dominants of both breeds and so a superior animal for beef purposes. When, however, these first crosses are bred together the bad recessive characters segregate and some individuals which have the bad characters (or dark face and horns) of both parent breeds are produced. This can also be illustrated by some experiments made by crossing dogs of different type (Fig. 11.6). For example, if the Boston terrier (top left) be taken as representing a dairy type of animal and the Dachsund (top right) be taken as representing a beef-type animal, then the result of the first cross (middle line) is a super animal. But if such crosses are interbred, segregation of types occurs in the second generation (bottom line).

However, the increase in variability on crossbreeding is inversely related to the number of genes controlling the character. Where there are few, as in the example involving colour and horns, increase in variability is great. When many genes are involved, as in the case of the quantitative, economic characters, the increase in variability of the F_2 over the F_1 is negligible and there is no reason to shrink from breeding from crossbreds.

Organized crossbreeding

There are two main reasons for the organized crossbreeding of domestic animals for commercial purposes. The first is to obtain good commercial returns from animals culled from pure breeds which are adapted to special environments (e.g. hills) or special industries (e.g. milk production). In the animal industry there is frequently a 'stratification of types', both in time and in space, from poor low-priced land to good high-priced land. Cross-breeding is practised to bridge the gap between these extremes economically. For example, in poor hilly country hardiness and suitability to their condition of life are the main considerations for the economic success of sheep; here is the place of the Blackface breed in Scotland. The cull ewes drafted from the hill flocks of this breed are taken to slightly better land where they are mated with rams of the crossing breeds such as the Border Leicester which are of larger size and are commonly regarded as being more fertile and having a higher milk yield. The males of this cross, the Greyface, are reared for slaughter, and the females are sent to the good

Fig. 11.5 Inheritance of characters in crosses between Aberdeen-Angus and Hereford cattle. Recessive characters like red colour, horns and dark face disappear in the first cross (*middle row*) which is quite uniform in appearance, but they reappear, when these first-cross animals are bred together, in the proportions shown by the numbers given (*bottom row*).

Fig. 11.6 The Boston terrier, with short round head and deformed tail end of vertebral column, crossed with long-headed long-tailed dachshund with short bent legs. The first-generation hybrids have long heads, long bodies and tails, and short bent legs. The second-generation hybrids exhibit various combinations of short heads and short legs and long heads and long legs. (Stockard, C. R. (1931). *The Physical Basis of Personality*. W. W. Norton Co., New York.)

pasture land in the north of England, where their progeny by the rams of the specialized meat breeds such as the Suffolk and Hampshire Down are sold as fat lambs. Some examples of suitable crossbreeding systems in sheep are shown in Table 11.1.

Table 11.1 Example of systematic crossbreeding in sheep to show how by using different crosses in succession the animal is fitted to a better nutritional environment (*left*, poor grass; *right*, good grass) where higher fertility and better mutton qualities are essential. (Hammond, J. (1947). *Suffolk Sheep Society Yearbook*. Ipswich.)

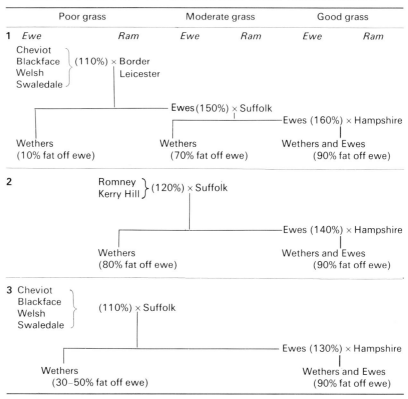

The figures in brackets (110%) after the ewes are an estimate of the average number of lambs born per 100 ewes mated. Naturally in individual cases these will vary widely owing to different circumstances such as whether the ewes were 'flushed' or not before the tup was put in, and whether they were kept in small flocks and on fresh grassland (higher fertility) or in large flocks and on old pasture (lower fertility). The figures give a general average for all conditions.

The figures in brackets (10% fat off ewe) after the wethers are an estimate of the average percentage of the lambs which can be got fat off the ewe without extra feeding on rape, sugar beet tops, etc., before they are fit for the butcher. These percentages are liable to vary widely in individual cases owing to different circumstances; for example, the 70% for the Suffolk × Border Leicester–Cheviot will include a high proportion of twin lambs, whereas the 80% for the Suffolk × Romney ewes will include a high proportion of single lambs. With better grassland (long leys) and with steaming up of the ewes for six weeks before lambing, higher results would be obtained than when this was not done. Lambs sold off the ewe in northern areas are taken to include those finished on aftermath.

A similar stratification holds in Australia. The cull Merino ewes (toothless grannies) from the pastoral areas, plus some cull Merino '2-tooths', are taken to the wheat belt and crossed with Border Leicester rams, their female progeny going to the subterranean-clover or irrigated areas to be mated with Dorset Horn or Polled Dorset (or sometimes Ryeland or Southdown) rams for the production of fat lamb. The qualities of hardiness, good fertility and good milk supply are brought in from the ewe, while good mutton qualities are brought in from the ram. Likewise, in New Zealand the cull Romney ewes from the hills are mated to Southdown rams to produce lambs which are sold from the low-country pastures as fat lamb.

The utilization of cull dairy cows for breeding beef animals is discussed on pp. 287, 288.

The second reason for crossbreeding is to exploit hybrid vigour by crossing two lines, strains or breeds and thus obtain a strong, vigorous, high-producing animal for commercial production in species, such as pigs and poultry, in which the breeding stock can easily be replaced. Thus in breeding fowls first crosses are made between two different breeds or strains. These are mated to a third breed or strain or to another first cross to produce vigorous second-cross birds for egg or meat production. A refinement of this crossbreeding for egg production is to inbreed the parent strains before crossing, as has been successfully done with maize. Many inbred lines are produced and many crosses made between them in pairs. Only the best crosses are exploited commercially, i.e. those lines which 'nick' best on crossing are retained.

While the first signs of lack of vigour on inbreeding are seen in the mothering qualities of the dam, so in reverse the most striking effects of crossbreeding are the increase in the fertility and milk yield of the dam and in the rate of growth of the young. Crossbreeding for meat purposes is based partly on the fact that good growth rate and early maturity depend on a good milk supply. This is particularly so with beef cattle, where many of the pedigree bulls are reared on nurse cows and so are not selected for quick growth based on milk supply from their own dam. Hence in Britain commercial beef herds usually consist of first-cross cows mated back to a bull of one or other of the parent breeds or to one of a third breed. Experiments in U.S.A. have shown that when Hereford cows were mated with Shorthorn bulls and their offspring put to Aberdeen-Angus bulls the calves weighed 212 kg at weaning as compared with an average of 176 kg for the pure breeds, and on similar feeds gave a final weight of 469 kg as compared with 414 kg for the pure breeds (Knapp et al., 1949). Similar results have been found in a large comparison involving temperate and tropical cattle and large breeds such as the Charolais and Chiaina, conducted at the Roman L. Hruska U.S. Meat Animal Research Center, Nebraska.

When British beef breeds are crossed the average of the two reciprocal crosses is better than the average of the two parent breeds by about 5 per cent for growth and viability and about 8 per cent for cow fertility (Mason, 1966). When British breeds are crossed with Zebus the amount of hybrid vigour is two or three times as great.

The same thing applies to pig breeding. If full advantage is to be made of hybrid vigour, crossbred females should be used as breeders. Table 11.2 shows the percentage advantages obtained from the first-cross litters and from litters of the first-cross sows by a boar of a third breed over those of the pure breeds concerned for different characters. It is evident that in birthweight per litter, number of pigs born and weaned, and weight at weaning, there is great advantage in using the first-cross sow. In many parts of the world pig production for pork and bacon is based on crosses of two, three or four strains or breeds. As with poultry the identity of breeds has been lost in most of the crossbreds used. Breeds of pigs such as

Table 11.2 Percentage advantages of pigs in first-cross litters over purebreds as compared with those bred from first-cross sows mated with a boar of another breed. (Winters, L. M., Kiser, O. M., Jordan, P. S. and Peters, W. H. (1935). *Bulletin of the Minnesota Agricultural Experiment Station*, 320.)

	First cross from purebred sow	Second cross from first-cross sow
Birth weight per live pig	1.96	0.39
Birth weight per litter of live pigs	13.39	20.65
Number of live pigs per litter	11.22	20.19
Total number of pigs per litter	4.04	8.62
Number of pigs weaned per litter	5.87	36.22
Litter weight at weaning	24.84	60.76
Saving in feed	2.99	3.85
Saving in time to reach 100 kg	8.67	8.63

Middle White, Tamworth, Berkshire and Welsh have declined to minor importance and most crossbreds are based on Large White, Landrace, with introductions of Hampshire, Duroc and the Belgian Pietrain.

In general, to obtain the full benefits from hybrid vigour it is essential to exploit the increased fertility and mothering ability of the crossbred females. This may be done by mating them to a third breed and slaughtering all the offspring for meat as described above, or by using a criss-crossing system, i.e. using sires of the two breeds in alternate generations. This retains hybrid vigour (or at least two-thirds of it) but is simpler in that only one kind of breeding female is kept instead of two—one pure and one crossbred.

Replacement crossing (or grading up) and the formation of new breeds also involve breeding from crossbred females, but after the first cross, do not efficiently exploit hybrid vigour.

Grading up

Replacement crossing enables us to improve low-producing stock rapidly by mating them, and their progeny in several successive generations, with purebred males of a highly productive strain. This process is known as grading up (Figs 11.7 and 11.8). In the course of about four generations

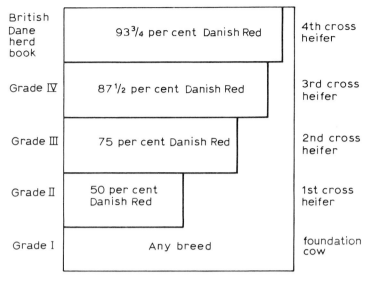

Fig. 11.7 Diagram showing how the grading-up scheme of the Red Poll Cattle Society works to produce a purebred British Dane heifer by using a Danish Red or British Dane bull in four successive generations. It is the four bulls used in succession in a herd rather than the foundation cows that will determine production level.

bred in this way one can obtain animals which will be indistinguishable in appearance and performance and will breed almost as well as the purebred animals to which they were graded. This is especially so if the males used for grading purposes come in each generation from the same strain, or the grade females are inbred in successive generations to the same purebred male. This method is now being used to replace one breed by another (Fig. 11.7) or to introduce the polled character into horned breeds of cattle (see Fig. 9.9).

Formation of new breeds

With plants the production of new varieties has been brought about by crossing two strains, each with some desirable character which it was desired to combine. The same system is used to form new breeds of animals. Where the breeds crossed do not differ materially in the unit factor characters or fancy points, it is easy to establish new breeds which have the mean values of their commercial characters lying at any point between those of the two parents crossed, according to the blood of each that is put in. As examples of this the Halfbred sheep (between Border Leicester and Cheviot), the Corriedale (a halfbred between Lincoln and Merino) and the Polwarth (a back cross of the Corriedale to the Merino) may be mentioned. However, it will be realized from what has been said above that, where the breeds crossed differ materially in the unit-factor segregating characters

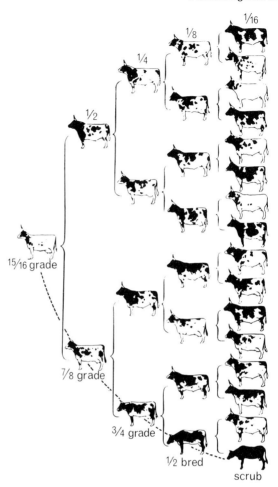

Fig. 11.8 Diagram to illustrate the system of grading-up a scrub cow (lower right) with purebred Ayrshire bulls. The figures on the top line show the fraction of the individual's genes most likely to come from each ancestor in the corresponding generation. (Modified from Finlay, G. F. (1925). *Cattle Breeding*. Oliver & Boyd, Edinburgh.)

(such as colour, horns and patterns), a large number of different types may arise when the second generation is bred. To eliminate the unwanted ones rigorous selection is necessary. Therefore large numbers of animals to choose from is an essential starting point in any programme of breed transformation.

In the U.S.A., a new red bacon pig, the Minnesota No. 1, has been produced by crossing the white Danish Landrace having good bacon qualities with the red Canadian Tamworth. Boars which had good bacon qualities and red colour in the second generation were progeny tested and those which bred true were used to form the new breed. Much the same

Fig. 11.9   Mature Santa Gertrudis bull. (Courtesy of Santa Gertrudis Breeders International, Kingsville, Texas.)

process was followed in forming a new breed of cattle, the Santa Gertrudis, to combine the beef qualities of the Shorthorn with the heat-resisting qualities of Zebu cattle (Fig. 11.9). A bull was found by progeny testing which bred true for the characters desired and he was linebred to form the new breed. Experiments are now being made in India and Australia to produce, in the same way, a new dairy breed of cattle suitable for the tropics by crossing European dairy breeds with the Indian Zebu.

In Britain the newest breeds are the Luing cattle (from Beef Shorthorn × Highland) and the Colbred sheep (from Clun Forest, East Friesian, Border Leicester and Dorset Horn).

References

KING, J. W. B. (1967). Pig breeding research. *Report of the Proceedings of the 9th International Congress of Animal Production, Edinburgh, 1966*, p. 9.

KNAPP, B. JR., BAKER, A. L. and CLARK, R. T. (1949). Crossbred beef cattle for the northern Great Plains. *Circular of the United States Department of Agriculture*, No. 810.

MASON, I. L. (1966). Hybrid vigour in beef cattle. *Animal Breeding Abstracts*, **34**, 453.

Further reading

DICKERSON, G. E. (1973). Inbreeding and heterosis in animals. Proceedings of the Animal Breeding and Genetics Symposium in honour of Dr Jay L. Lush.

American Society of Animal Science and *American Dairy Science Association*, pp. 54-77.

PEARSON DE VACCARO, L. (1973). Some aspects of the performance of European purebred and crossbred dairy cattle in the tropics. 1. Reproductive efficiency in females. *Animal Breeding Abstracts*, **41**, 571.

PEARSON DE VACCARO, L. (1974). Some aspects of the performance of European purebred and crossbred dairy cattle in the tropics. 2. Mortality and culling rates. *Animal Breeding Abstracts*, **42**, 93.

12 Breeding for production in the different species

In the past the showyard has played a large part in moulding our breeds of farm animals. Criticisms of this method have been made on the basis that selection has tended to follow appearance rather than commercial qualities. However, the changes in body conformation and composition, in wool quality and yield, and in milk let-down and yield which were effected by the methods of breeders in the 18th and 19th centuries, should not be underestimated. Though these agricultural improvers, such as Bakewell and the Collings brothers in the U.K., kept few records and used fairly simple selection methods there is no doubting the beneficial effect they had on the performance of farm livestock.

During the last thirty years there have been substantial developments in the selection methods used, in the size of populations available for selection and in the ease of collecting, storing and analysing large numbers of performance records. The appearance of an animal on the farm or in a showyard is still an important element in determining whether it will be chosen to be a parent of the next generation. However, its performance and that of its close relations for characters of commercial significance are now of overriding importance in the selection process. A high performing animal will be used as a parent if it also has a good appearance but a poor performing animal will be culled however good its appearance.

Selection objectives

The first thing to do when starting a selection programme is to decide objectives (characters to be selected) and this can be much more difficult than is often supposed. It is necessary to define accurately those characters which are commercially important now and those which will be important 10 to 15 years hence, for the animals from the selection programme will not have much effect on commercial production until then. Having defined characters in this way, difficulties arise because there are usually too many of them for selection to be effective and some of them may not be capable of direct measurement without considerable expense or without killing the animal. These problems can be illustrated by reference to beef cattle. Table 12.1 shows some characters which might be subjected to improvement by selection. This list exemplifies some of the difficulties in choosing selection

Table 12.1   Characters which might be included in selection objectives for improving
cattle for beef.

Growth rate	Carcase quality
Feed consumption	killing out percentage
Feed conversion	lean percentage
Resistance to disease	fat percentage
Behaviour	meat colour
docility	meat tenderness
	keeping quality
	eye muscle area

objectives. Growth rate can be measured easily, relatively accurately and cheaply. Feed conversion is more difficult. The feed intake of individual animals can only be obtained by recourse to individual penning or to the use of electronically controlled gates, which allow each animal access to its own individual feed trough even though the animals are housed in groups. Both methods are unlike conditions found on commercial units, so that feed intake measurements obtained under test conditions may be unrepresentative of those on farms. Carcase quality measurements can be obtained effectively only by slaughtering the animal, though new techniques for measuring carcases in live animals will reduce considerably the number which have to be slaughtered to measure carcase characters (see p. 264). Thus it is necessary to store semen from potential breeding males before slaughter, or to rely on the records of carcase quality of relatives, to be able to select for these carcase characters. The cost of measuring carcase traits is high. Resistance to disease is very difficult to select for because of the problems associated with offering an equal and realistic field challenge to each animal. Behaviour characters are still largely described in qualitative terms which do not make selection easy. Reproductive performance presents special problems referred to below.

This example of choosing characters for selection is not atypical and it can be appreciated why selection in beef cattle has been concerned largely with growth rate and appearance. It is well to remember that the more characters included for selection the less the likely progress for each of them.

Selection to suit farm environments

As part of planning a selection programme it is necessary to determine the possible range of farm production systems in which the animals will be used, and whether different selection schemes will be needed to breed specialized strains for each system. Selection will usually be under the system giving maximum expression to the character being selected. In general, strains of livestock are extremely adaptable to a range of conditions. Only if extreme environments or very different market cireumstances are considered is it found necessary to contemplate more than one selection programme. Generally it is simpler and more profitable to eliminate the extremes of environment by altering the production system than to multiply the number of selection programmes.

The structure of animal breeding programmes

Until about 1950, most animal breeding was the responsibility of individual herd owners. However, the more important programmes are now the responsibility of several other types of organization. These include government organizations, breeder cooperatives and breeding companies. The latter having started to select one animal species, mainly poultry (or even plants), have in several cases diversified to breed several species. It is interesting to discuss the reasons for these developments. In the first place a very large population is not essential for genetic progress, so that there is nothing to deter one individual herd owner from continuing a constructive breeding programme with the animals at his disposal. However, the cost of testing animals, and of culling animals in order to apply some worthwhile selection pressure, is high in relation to the cost of production. Also there is no doubt that the larger the population under selection the greater the chance of success. The cost of the programme has to be more than recouped if the breeder is to remain in business. It may be several years between the start of the selection programme and the time when the benefits in terms of improved animal performance can be recouped by the breeder, by the sale of stock to commercial producers. The breeder has to carry the cost for this period. The more progeny produced from improved stock, the lower the cost of the programme as a charge on each animal produced. Thus multiplication, marketing, and indeed organization, play extremely important roles in the financial success and long term viability of a breeding programme. These programmes are dependent on fast data processing equipment, on technical advice and on continuing access to new genetic developments, all expensive facilities. However, since the principles are similar irrespective of the species, it is not difficult to run simultaneously several breeding schemes, for different species, using the same technical resources. Savings in marketing costs can also be achieved. Apart from all these factors militating in favour of the larger organization for animal breeding there is also pressure from commercial producers and processors. These interests, which also have grown and are growing in scale of operation, require less variation in the animals used. The numbers and quality required can be supplied only by breeders whose multiplication can offer, for example, thousands of gilts per year of the same breeding. It is not only the needs of selection which favour the large breeder but also the economies of scale of operation, multiplication and marketing, coupled with producer requirements.

Fertility

Fertility, or the number of young produced at birth, is a complex character and depends not only on the number of ova shed by the female and on the amount of sperm production by the male, but also on the number of the fertilized eggs which develop properly up to the time of birth. Experiments have been made on rabbits (Table 12.2) to determine how fertility is inherited. Two low-fertility inbred strains were crossed together; these produced different numbers of eggs, one strain averaging 6 and the other 11, but much the same number at birth (average 3 and 5 respectively). The first-generation does produced double the number of young (7) given by the parent strains. This was because the two strains crossed were of low fertility for different reasons; one strain because of the small number of eggs shed (6) and the other strain because, although a large number of eggs were shed (11) and fertilized, many of the embryos atrophied *in utero* before birth (Fig. 12.1). The number of eggs shed behaves as a multiple-factor or blending character and is just above intermediate (10) between the two parents.

Table 12.2   Inheritance of fertility in crosses and backcrosses between two inbred lines of rabbits (H and F); results of post mortem at the 29th day of pregnancy. (Hammond, J. (1932). *Proceedings of the international Genetics Congress, Ithaca, New York:* (1934). *Report of the 6th Rabbit Conference, Harper Adams College.*)

Strain or cross	No. of eggs shed	No. of embryos		Atrophic embryos per 100 normals
		Normal	Atrophic	
H	11.1	4.8	4.0	83.0
F	5.8	3.2	0.5	15.8
W (H × F)	10.4	6.8	1.2	17.1
U (W × W)	9.0	5.9	1.6	27.1
S (W × F)	8.2	6.7	0.4	6.0
Z (W × H)	10.3	5.7	2.8	49.1
J (Z × H)	11.3	4.4	4.4	100.0

Foetal atrophy in rabbits is a maternal character, that is, it is due to the lack of some secretion in the mother's uterus and not to the genetic composition of the embryos, for females of the foetal-atrophy strain have just as much foetal atrophy when mated to unrelated males as they do with males of their own strain.

In the pig, foetal atrophy is frequent, but there is every reason to suppose that it is inherited in the same way as in rabbits, for it appears as a maternal recessive character on inbreeding. In the sheep on the other hand there is little foetal atrophy, so the number of young born depends mainly on the number of eggs shed. Breeds differ considerably in this respect.

Experiments in New Zealand have shown that it is possible to increase fertility in sheep by selection. Starting in 1948 a Romney flock was divided into three groups; in one group selection was made for high fertility, in one no selection was made, and in one selection was made for low fertility. During 1948–51 the fertility percentage (lambs born per 100 ewes lambing) was 119 for the high-fertility group, 116 for the no-selection group and 100 for the low-fertility group. During 1956–9 the average fertility percentages were 137, 118 and 119 respectively and for 1967 they had

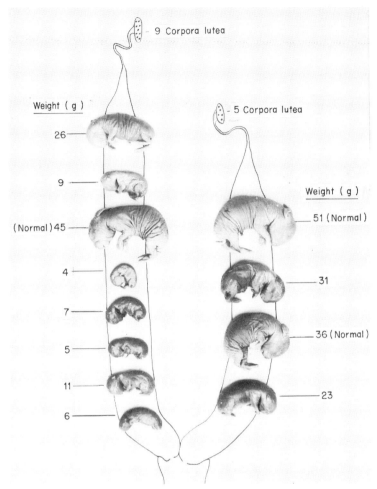

Fig. 12.1 Foetal atrophy in the rabbit. Embryos taken from the uterus of a doe which was from an inbred line breeding true for this character, pregnant 29 days; only 3 normal embryos out of 12 which started development were alive. (Hammond, J. (1928). *Zuchtungskunde*, **3**, 523.)

advanced to 152, 119 and 108. The slow progress in the early years may have been due to absence of lifetime records of ewes from which to select rams. The rams used in succession on the flock would have far greater influence than the culling of ewes for low fertility. In the initial stages of selection before lifetime records have become available, mating the ewe lambs will, age and feeding conditions being equal, pick out those which have a high potential for fertility (Wallace, 1964; New Zealand, 1969).

Crossing with a prolific breed, however, is a quicker method of increasing fertility in sheep. The Finnish Landrace breed (Fig. 12.2) which averages 2 lambs per litter for one-year-old ewes and 3 (or even more) at

Fig. 12.2 Finnish Landrace ewe with triplets on Animal Breeding Research Organization farm near Edinburgh. Note the characteristic short bare tail of the ewe. (Courtesy of *Farmers Weekly*.)

older ages, is now being tried out in several countries. As expected, when crossed with local breeds, fertility is intermediate.

There seems to be no reason why twinning strains of cattle should not also be produced by selection. Possibly the best way of doing this would be to collect cows which had been recorded in the herdbooks as producing several sets of twins. Bulls to mate to those cows should be obtained from those cows which had produced several sets of twins in their lifetime, and by choosing bulls whose daughters have a high record of producing twins.

The heritability of twinning rate is low, so selection would be a very slow process. However, by using superovulation and ova transplant techniques progress could be quite fast and might increase the incidence of twins from 2 per cent to 20 per cent in a few years.

The time of onset and length of the breeding season in sheep are inherited characters which behave as intermediates on crossing. In Dorset Horns on the average the breeding season begins in July and lasts for 12 cycles as compared with November and 7 cycles in the Welsh, while the first cross begins to breed in early October and has 10 cycles (see Fig. 5.1).

The present understanding of the physiological mechanisms which determine genetic difference in the reproductive performance of sheep has led to the suggestion by Land (see Fig. 12.3) that many aspects of reproductive activity in males and females are controlled, at least in part, by the same hormonal or metabolic variables. There is then the possibility that the reproductive or production performance of females and of a male's poten-

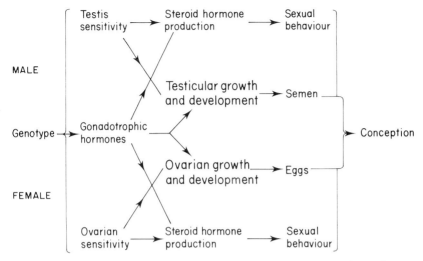

Fig. 12.3 The central role of the gonadotrophic hormones in the pathway from genotype to conception in males and females. (Land, R. B. (1974). *Animal Breeding Abstracts*, **42**, 155.)

tial offspring can be predicted from measurements of amounts of hormones or of substances resulting from hormone action in young lambs. Similar ideas are being explored in cattle and pigs.

Horses

Horses are bred for two main purposes, speed and draught. The natural evolution of the horse is for speed and this is associated, among other things, with increase in leg length (see Fig. 3.8). This can be seen when fossil forms are compared on similar cranium size (ear-eye length). In the development of the individual these changes are repeated during foetal life. The proportions of the Thoroughbred form a continuation of these series and follow the normal course of evolution. Draught power requires weight and thickness of muscle and bone, which is the natural development in postnatal life, and it is by the intensification of this process that heavy breeds of horses have been developed.

It would seem that to improve the horse for speed the amount of growth early in life must somehow or other be increased, while to improve the horse for draught the development of the body after birth must be greatly increased, possibly by rearing on a high plane of nutrition such as is done in the improvement of meat animals.

In addition to the skeletal structure and the conformation, especially of the hind leg, back and rump muscles, the temperament of the horse is very important for its value as an animal to work with man. Not only does the horse need to be docile and quick to respond to instruction but it also has to have staying power. A great deal of horse selection has been done on individual performance and from pedigree records, using ancestor performance to choose elite mares and stallions. Now the techniques of quantitative genetics are being applied to records on thousands of horses. Heritability estimates have been calculated for many traits including temperament, 0.23; gaits, 0.41; conformation, 0.23; soundness of feet and legs, 0.25 (Varo, 1965). The analysis has been extended to racing performance with a heritability of 0.35 (More O'Ferral and Cunningham, 1974). These values indicate that genetic improvement of the horse for speed and draught could be achieved quite quickly using methods similar to those for other farm animals. For gaits and racing ability individual selection is appropriate whereas for temperament, muscling, legs and feet progeny testing would give faster gain. Heritability estimates for the same character can vary quite markedly, depending on the population on which they are calculated and the method of calculation. Wherever possible, heritability should be measured on the population being selected.

Pigs

Pigs have been bred for three main purposes—lard, pork and bacon. Lard is now of much less importance, and all three products are being obtained from the same type of pig slaughtered at different weights. For pork production, some parts of the carcase are of more value than others, so that attempts have been made to increase the proportions of the more valuable parts. As the pig grows its proportions change; during the early foetal stages the head and neck are proportionally large and the loin small, but after birth, in our improved breeds, the proportions of the loin rapidly increase (see Fig. 6.13). When the unimproved pig, the wild boar, grows it does not change its proportions to anything like the same extent and remains, like the embryo, relatively heavy in the head and shoulders. Improvement has apparently come by selecting for increased loin development in the adult and the pushing back of this into an earlier stage of development, as occurs in natural evolution. That is, such differences tend to be inherited not at a corresponding age but at an earlier age, just as the embryonic forms tend to recapitulate the evolutionary history of the breed. Thus types of pig, such as the Berkshire and the Middle White, go through these changes in proportions more quickly, and are ready for slaughter earlier, than other types, such as the Large White (see Fig. 6.13). That is, the character we are breeding for in the early slaughter type is a rapid change in body proportions which makes them suitable for killing at low weights (see Fig. 6.15). As previously pointed out, rapid change in body proportions is only possible on high nutrition and so selection is best made under conditions of good nutrition.

With this rapid change in body proportions has come an increase in those tissues, such as fat, which normally develop late in life. This is an advantage when lard is the main objective, but since the sale for this has decreased, the Poland China breed in the U.S.A. has been turned into a breed (see Fig. 6.16) with more lean and less fat by increasing its size, and so delaying age to slaughter, as well as by changing the showyard type progressively.

Present market requirements are for much leaner carcases than in the past. There are some differences between breeds in carcase lean content, with the Hampshire and Pietrain having good reputations for this trait. Attempts to increase the proportion of lean and reduce the proportion of fat in pork and bacon pigs by selection and by feeding and management, have sometimes increased the incidence of other problems of meat quality. In particular, when pigs have been selected for blockiness in an attempt to increase muscle to bone ratios, there has been a marked increase in pale, soft, exudative (PSE) pork and also dark firm dry (DFD) pork. These defects are probably of different genetic and environmental origins. There appears to be little relationship within British breeds between shape and lean content or proportion of high priced joints. Selection for leanness has been done by reducing fatness with less emphasis on conformation. PSE or watery pork is meat in which the red colour disappears and the muscle has poor water holding capacity. The meat is undesirable because of the weight loss and because it is less suitable for curing. PSE pork is more common in pigs which are subject to stress such as caused by transport and herding just prior to slaughter. After slaughter such pigs have a rapid depletion of muscle glycogen and a low muscle pH which affects the water holding capacity. PSE condition in pigs has been shown to be closely associated with halothane-induced 'malignant hyperthermia' syndrome. Pigs can be tested early in life by exposure to halothane gas. Those which respond can be discarded, so saving expensive test facilities. Halothane sensitivity has been shown to be due to a single gene effect.

Boars tend to grow faster, have a better feed conversion and be leaner than castrates and gilts. There has therefore been much interest in leaving boars entire for pork and bacon production. However, meat processors are experiencing curing problems and excessive leanness with boars, also meat from a small proportion of boars contains an odour which can be detected, and is disliked by consumers. Research is being done to try to avoid the problems of boar odour.

In order to encourage the right system of feeding and management so that selection of pigs for commercial characters can be efficiently carried out, pig recording has been undertaken. The important records to keep are the number of pigs born and weaned to indicate the fertility and mothering qualities of the sows. In addition, the weight of the young pigs at 3 weeks old should be recorded to measure the milking qualities of the sows, for up to this time they are almost entirely dependent on the mother's milk. Weights of the young pigs can also be taken at 8 weeks of age, to measure not only the milk supply of the sow, but also the efficiency of the weaning and early rearing system adopted, because this is the critical stage in the life of the pig for obtaining length of side and efficiency of feed conversion. The age at which the pig reaches slaughter weight for pork or bacon, the feed consumed and the carcase quality are also important characters to be recorded. The heritabilities of some of these characters are shown in Table 12.3.

The chief means of improvement for growth and carcase characters is by progeny and performance tests, which in the past have been mainly carried out in testing stations but are now increasingly being done on breeders' own farms. Performance testing of boars (see p. 265) has the advantage that large numbers of boars can be tested and unsuitable ones eliminated before they produce offspring. The characters that can be accurately tested in the live pig are growth rate, feed conversion, and depth of backfat (by either the probe or ultrasonic—echo-sounding—method). The latter measures the depth by the time taken for the echo to come back from the junction of fat and lean. An accurate assessment of carcase quality, particularly carcase lean, requires that a proportion of pigs be slaughtered and dissected, but this is no longer essential.

Table 12.3 Heritabilities of important characters in two breeds of pigs in the U.K. (From Smith, C., and Ross, G. J. S. (1965). *Animal Production*, **7**, 291.)

Trait	Probable values	
	Large white	Landrace
Growth rate (gain per day)	0.41	0.41
Feed conversion (feed/carcase gain)	0.58	0.48
Dressing out (%)	0.40	0.26
Loin length (mm)	0.46	0.39
Average backfat (mm)	0.66	0.74
Eye muscle area	0.35	0.49

Reproductive characters such as litter size generally have low heritabilities and are very difficult to improve by within herd selection. Most of the gains in reproduction in recent years have therefore been obtained by crossbreeding (see p. 250) and by using hybrid sows for commercial pig production. There may also be some advantages, such as enhanced libido, from using hybrid boars; but if the types included in crossbreeding are very different for growth and carcase characters the offspring from matings of hybrid sows and hybrid boars may be very variable for carcase quality.

In the U.K. in 1971 the Meat and Livestock Commission (MLC) re-organized their Pig Improvement Scheme which had been operating since 1966. The revised scheme has four categories of participant, though they tend to overlap: (1) The nucleus herd; (2) the reserve nucleus herd; (3) the breeding company; and (4) the nucleus multiplier herd. The most efficient way to use the testing capacity of the MLC central stations is to make them available to a few herds at the summit of the breeding pyramid (see p. 245). These herds are officially designated 'nucleus'; they were chosen by an independent panel of breeders and scientists according to breeding records and management standards. They are encouraged to progeny test all their boars annually, to breed only from boars from nucleus herds, and to maintain a short generation interval. Gilt performance is tested on the farm. Reserve nucleus herds are those aspiring to become nucleus herds. After panel approval reserve nucleus herds test their pigs at MLC stations for one year at least. On the basis of the results the panel decide whether the reserve herd can be upgraded to nucleus category.

There are an increasing number of breeding companies which carry out gilt and boar testing on their own farms, but also have access for boar testing to the MLC test stations, to an extent determined by the independent panel. These companies may sell purebred and hybrid boars and gilts and may make use of breeding stock from the nucleus herds. The companies using the MLC stations also take part in the MLC Commercial Pig Evaluation Test, to compare the performance of commercial breeding stock and slaughter pigs sold by the companies.

The nucleus multiplier herds buy gilts and tested boars from the nucleus herds and multiply these for sale of purebred or crossbred stock to commercial herds. Some selection of gilts may be made in nucleus multiplier herds provided the herd uses the MLC on-farm ultrasonic testing service for fat measurement. All types of herds and companies using MLC facilities must be members of the Government Pig Health Scheme.

The MLC boar performance test, or combined test as it is known, measures a young boar's potential to sire above average progeny. From any one litter of a nucleus or company herd, a test group consists of two boars on performance test and a castrate and a gilt for slaughter. During 1977–78 about 3000 boars were tested under standard feeding and management conditions in this way. The breeding value of the boars is assessed from their own performance and that of their littermates. The pigs enter

Table 12.4 An example of a report of a Meat and Livestock Commission (MLC) combined test. The points score for an individual boar is compared to the average of all boars in the same test, which is set at 50 each for economy of production and carcase quality. Information on individual characters are also compared to contemporary boar averages.

Points score				
	Ear no. of boar		Average of both boars	Contemporary average
	7494	7495		
Economy of production	64	54	59	50
Carcase quality	59	65	62	50
Total	123	119	121	100

Estimated genetic superiority of individual characters

Character	Ear no. of boar	
	7494	7495
Daily gain (g)	6 (B)	2 (B)
Food conversion	0.036 (B)	0.026 (B)
Killing out %	0.05 (B)	0.12 (W)
Trimming %	0.01 (W)	0.03 (W)
% lean in side	0.33 (B)	0.57 (B)
Eye muscle area (cm²)	0.21 (B)	0.22 (B)

B = better; AVE = average; W = worse

the testing station when their individual weights are between 18 and 25 kg and the test starts when each boar weighs 27 kg and the sibs total 54 kg. The test is completed when each boar weighs 91 kg and the sibs' total weight is 165 kg, at which time the sibs are slaughtered. Detailed carcase measurements are taken and feed consumption, liveweight gain and ultrasonically measured fat depths are recorded on each boar. This information is combined with feed consumption and growth records from the sibs, plus their carcase measurements, and expressed in a selection index as a points score for each boar (Table 12.4). In order to allow for environmental differences between different stations and at different times of the year (see Fig. 12.4) the two scores are presented as contemporary comparisons, i.e. they are compared with the average of other boars tested at the same station at the same time. The average score is always maintained at 100 points, although the actual performance of boars in the testing stations is improving year by year. Boars scoring below 90 points are slaughtered and boars over 90 points are offered back to the breeder. Breeders are advised to use boars with the highest scores to make the most genetic progress. Boars scoring more than 150 points are usually retained in the nucleus herd or go to an AI stud for widespread use (Table 12.5).

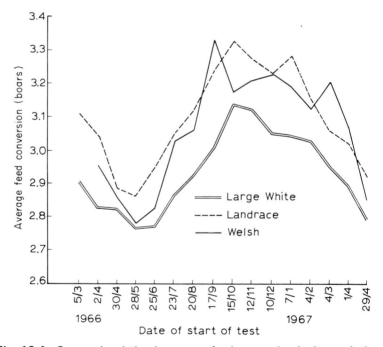

Fig. 12.4 Seasonal variation in average feed conversion by boars. As boars are housed in huts in open buildings, feed conversion differs considerably between summer and winter. Most of the seasonal effect can be removed by comparing each boar against its contemporaries. The diagram shows that seasonal changes are similar for different breeds. (*P.I.D.A. Accreditation Yearbook; Second Register of Elite and Accredited Herds 1967–68.*)

Table 12.5 The fate of boars tested in MLC central performance test stations during the twelve months ended September (1978). (Meat and Livestock Commission, Milton Keynes.)

	Number	Percentage
Returned for use in nucleus herd or company herd	1186	36.6
Slaughtered by the breeder	420	12.9
Casualty after test	16	0.5
Incomplete test	88	2.7
Slaughtered–score below 90	1534	47.3
Total	3244	100.0

The genetic progress made by the pig improvement scheme and the difference in performance between scheme pigs and those from ordinary commercial herds is monitored regularly by MLC. Groups of pigs from specially maintained Large White and Landrace control herds (p. 224) are tested alongside improvement scheme pigs in MLC test stations. The difference in performance between these two sources of stock is becoming larger each year and provides an estimate of annual genetic progress (Table 12.6).

Table 12.6 Genetic change in pigs tested by the Meat and Livestock Commission over eight years to 1978. (Jones, D. (1980). Meat and Livestock Commission, Milton Keynes.)

Character included in selection index	Rate of genetic change per year
Feed conversion ratio	−0.0269
Daily gain (g)	+4.91
Eye muscle area (cm²)	+0.266
Killing out %	+0.109
Trimming %	+0.082
% lean in side	+0.683
Other characters	
Daily feed intake (g)	−7.98
'C' fat depth (mm)	−0.562
'K' fat depth (mm)	−0.692

The MLC Commercial Pig Evaluation Test checks the performance of gilts and boars sold by breeding companies. Thirty hybrid gilts and seven boars are purchased on a random basis from each of nine companies and housed under commercial conditions at one test farm. The test is based on the results of two litters per female; approximately 200 of the progeny of each company are fed to three different weights on two feed treatments— restricted and ad libitum. Three slaughter weights selected are 61 kg, 91 kg and 118 kg. The reproductive performance is fully recorded to provide information on number of piglets born and reared, piglet mortality and weaning weights. Feed conversion ratio, daily liveweight gain and normal carcase classification measurements are recorded for the slaughter pigs. A proportion of pigs at all weights is fully dissected. Company pigs are compared with each other, with either purebred pigs or commercial crossbred gilts, and with pigs from the control herds.

In addition to these testing and selection programmes, MLC have made available since 1971 a pig carcase classification scheme. Now some 78 per cent of all pigs slaughtered (excluding sows and mature boars) are being classified. The main purpose of the information is to allocate carcases to appropriate markets and for buying purposes, but it can also be sent to the pig producer. In this way the producer can choose the type of pig and the method of feeding and management. The classification scheme is based on five factors:

cold carcase weight;
backfat thickness;
visual assessment applied to certain carcases—scraggy, deformed, blemished, pigmented, coarse skinned, soft fat, pale muscle, partially condemned—these are classed as Z;
length;
conformation—carcases with poor conformation (particularly of the leg but not poor enough to be scraggy) are classed as C.

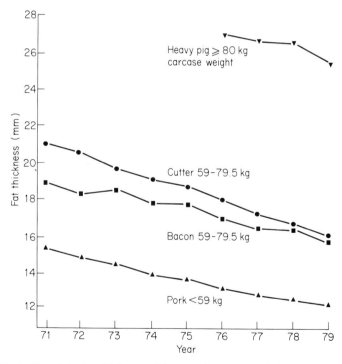

Fig. 12.5 Trends in the thickness of fat over the eye muscle in different market types of pig following the introduction of Meat and Livestock Commission (MLC) carcase classification scheme. (Kempster, J. A. (1980). Meat and Livestock Commission, Milton Keynes, U.K.)

Measurements of thickness of subcutaneous fat over the eye muscle, at the level of the last rib, are made with an optical probe (intrascope) either $6\frac{1}{2}$ cm, or both $4\frac{1}{2}$ and 8 cm, from the dorsal midline (the former usually for bacon, the latter for other types of carcase).

The effect of the classification scheme on the quality of pigs being offered for slaughter has improved markedly in the eight years since it was introduced (Fig. 12.5), even though carcase weights have remained almost constant.

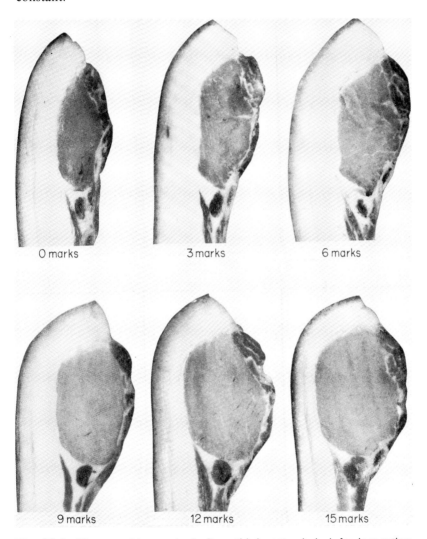

O marks   3 marks   6 marks

9 marks   12 marks   15 marks

Fig. 12.6 Photographic standards (two-thirds natural size) for inspection judging of the size and shape of the eye of lean in cured bacon sides cut at the level of the last rib. Maximum marks 15. (National Pig Breeders Association (1959). Advertisement Committee for Pig Carcase Competitions. Watford.)

Competition carcase tests also to some extent serve a similar purpose and were very effective in England, New Zealand and Australia in stimulating the improvement of bacon and pork pigs. In such competitions points are given so that competitors can see why their carcases do not win and so can take steps to improve them. In order to standardize inspection judging at different shows and by different judges, photographic scales have been produced; for example Fig. 12.6 shows the standards for size and shape of eye of lean. In the actual scale used the photographs are reproduced in natural size. These carcase competitions have received great impetus from the fact that bacon pig carcases are sold on grade.

Sheep

Sheep are kept either for meat production or wool production or a combination of the two products. The heritabilities of the important characters are given in Table 12.7. Selection programmes for sheep have been limited in the United Kingdom; in New Zealand and Australia more comprehensive schemes have been developed for selecting for growth and carcase character, wool characters and for number of lambs born. In Australasia large populations on any one farm have been available for selection, but even larger populations have been obtained by the organization of nucleus breeding schemes. In such cases several breeders collaborate and select their best animals for transfer to an elite nucleus. Young males are supplied from the nucleus for progeny testing in the collaborating flocks and the best of them are returned to the nucleus to breed more young males for testing. In some cases the best females from the collaborator flocks are also supplied to the nucleus, but many of the nucleus flock female replacements are bred in the flock from the very best progeny tested males mated to nucleus flock females.

Sheep recording is not easy to organize because of the extensive grazing conditions and limited handling which most sheep experience. However, in 1968 MLC in the U.K. started a pilot sheep-recording scheme with the aims of discovering the problems of collecting and interpreting flock records, of establishing a series of recorded flocks throughout the country which would serve as a basis for developing local study groups, and of accumulating information for the design of possible improvement programmes. The recording scheme involves both commercial flock records and pedigree records. Several hundred flocks have participated and valuable information on comparative breed and crossbreed performance, albeit under different circumstances of feeding and management, has been obtained (for example Table 12.8).

The main object in breeding for improved lamb production, apart from improved lambing percentage and ewe milk yield, is to increase the proportions of the later-maturing parts of the body (e.g. loin) and the lean content of the carcase. The wild Mouflon undergoes changes in body proportions as it grows up (see Fig. 5.19), but in our improved breeds these have been extended in the adults, and pushed forward to an earlier stage

Table 12.7 Range of heritability estimates for various characters in British breeds of sheep. (MLC Sheep Improvement Scientific Study Group Report, October 1972. Milton Keynes.)

Trait	Probable values
A. *Feed intake of ewe*	
1. Mature body size	0.15–0.55
2. Efficiency of feed conversion	Not known
B. *Reproductive traits of ewe*	
1. Ewe mortality rate	Not known, probably below 0.05
2. Date of first oestrus (in particular in those breeds in which out-of-season lambing is important	0.25–0.35
3. Proportion of ewes lambing among those put to the ram	0.00–0.10
4. Litter size at birth	0.10–0.20
5. Lamb mortality before weaning	0.00–0.05
6. Milk yield of the ewe	0.10–0.20
C. *Characteristics of the lamb*	
1. Viability	0.00–0.05
2. Growth rate	0.10–0.30
3. Feed intake	Not known
4. Feed conversion efficiency	Not known
5. Carcase composition and conformation	0.25–0.35
D. *Fleece*	
1. Fleece weight	0.30–0.45
2. Fleece quality	0.40–0.70

of development, so that the body proportions of the adult Mouflon ewe are very similar to those of an improved Suffolk at 3 months old. This has apparently been done by selection in an environment of high nutrition which develops the proportions of the body to the fullest extent. All the improved early-maturing Down breeds of sheep, for example, have been produced under conditions where cultivated crops are grown and folded so that a uniform state of high plane nutrition is maintained. For fat-lamb production where the carcase has to be marketed at a low weight (15–17 kg), the early-developing parts of the body must be reduced so that more of the nutrition available goes to the later-developing parts; this, for example, means a reduction in cannon bone length and body size. Whilst it was thought that selection for early growth rate and carcase quality would have to be by progeny testing, recent research indicates that performance testing of ram lambs of Down breeds can be used to improve growth rate and feed conversion.

In order to enable the breeder to find out in what particular respect his animals are meeting, or not meeting, the requirements of the market, MLC have developed a sheep carcase classification scheme. It is based on the four factors weight, category (lamb, hogget, or sheep), fatness as determined by a visual appraisal of external fat development, and conformation, into four classes by visual appraisal of shape, taking into account carcase

Table 12.8 Summary of performance of commercial crossbred ewes in recorded flocks 1970/72. (MLC Sheep Improvement Scientific Study Group Report, October 1972, Milton Keynes.)

Cross	No. of flocks	No. of ewes	Bodyweight at mating (kg)	No. of productive ewes	Total lambs born	No. dead or died at birth	No. of lambs reared	Live lambs per ewe lambed
Masham	8	659	70.8	93	166	19	139	1.72
Greyface	6	396	65.0	92	167	9	149	1.83
Mule	16	1687	71.7	81	165	13	146	1.82
Welsh Halfbred	22	3789	57.1	93	134	8	121	1.44
Scottish Halfbred	21	3164	74.4	93	165	14	144	1.82
Suffolk × Scottish Halfbred	21	2975	78.0	89	151	14	131	1.70
Romney Halfbred	5	546	73.5	90	122	12	106	1.36

thickness and blockiness and fullness of the legs (Fig. 12.7). In assessing conformation fatness plays its part in influencing overall shape and no attempt is made to adjust specifically for fatness. Among carcases of similar weight and fatness, good conformation carcases tend to have greater lean meat thickness than those of poorer conformation.

Occasionally sheep occur which have yellow fat, a most undesirable character; like the similar condition in rabbits, this appears to be a simple recessive character.

Body weight tends to increase with increasing fleece weight. It should, therefore, be possible to increase meat production without detriment to wool production. Nevertheless, breeding for wool alone, as in the Merino, presents a simpler problem. For wool improvement, weight of fleece and fineness of wool fibre, among other things, are required. These are affected considerably by nutrition. Selection for fleece weight increases staple length and density (fibres per unit area) but it also increases skin wrinkles and fibre diameter. An increase in fibre diameter means a reduction in quality. Skin folds must be rigorously selected against because they lead to shearing difficulties and increased susceptibility to fly strike. Uniformity of fibre diameter over the body is said to be important but it is very difficult to measure and therefore to select for. In general, mass selection for fleece weight and wool quality gives more rapid improvement than progeny testing.

Dairy cattle

In the conversion of the protein and energy in feeding stuffs into animal produce suitable for human consumption there is loss of nutrients. This conversion to milk by the cow is on the average more efficient than for most of the other animal products (see Table 4.8). There is, however, a great difference between individuals in this efficiency of conversion; in low-yielding cows a much higher proportion of the feed intake is required for maintenance (see Fig. 8.5). Since the main expense is feed, this is one of the chief reasons for breeding for high milk and butterfat production.

The selection objectives for dairy cows depend on the method of payment for milk. Whilst there has been a major emphasis on milk yield until recently, there is now more importance attached to yield of milk components such as fat and protein. Some countries such as Holland have even adopted a negative payment for increased milk volume to deter the production of water.

The heritabilities of the important characters in dairy cattle are given in Table 12.9. Most selection schemes are designed to improve milk characters but little has yet been done to select for improved efficiency of feed conversion, except by increasing yield. The look of the animal, particularly with regard to mammary structure and the quality of feet and legs—very important to a grazing animal—are included in type classification schemes which are used as additional criteria to milk characters for selection of cows and of bulls via their daughters.

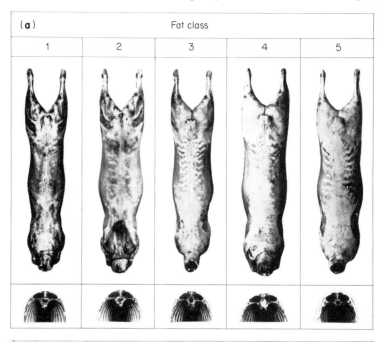

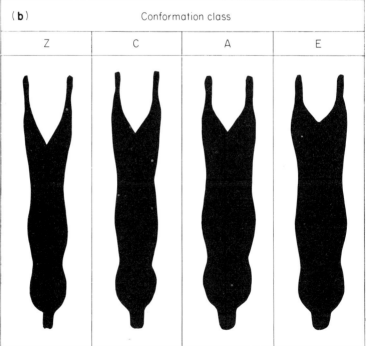

Fig. 12.7 MLC classification of lamb carcases. **(a)** For fatness—ranging from very lean **(1)** to very fat **(5)**; and **(b)** for conformation—very poor **(Z)**, poor **(C)**, average **(A)** and very good **(E)**.

Table 12.9   Heritabilities of the more important characters in dairy cattle.

Trait	Probable values
Milk yield	0.25–0.35
Fat yield	0.25–0.35
Protein yield	0.25–0.35
Fat percentage	0.45–0.55
Protein percentage	0.45–0.55
Total milking time	0.25–0.35
Mastitis resistance	0.20–0.35
Udder shape	0.10–0.20
Calving difficulty	0.00–0.25
Body weight	0.25–0.40

In the past the main method of improving milk yields was by milk recording and culling of the low yielders. This was wasteful because the culling is done after loss has been sustained through low milk production. The overhead food costs of rearing a heifer to milking age make the cost of milk production high if she has to be eliminated from the herd through low yield or disease before she has reached the age ($7\frac{1}{2}$ years) of maximum production (see Table 12.10).

Genetically, the bull must bear half the responsibility for the production of low-yielding heifers, so that better selection of the bulls used in dairy herds is a more effective and economical means of improving milk and butterfat yields than is culling the low-yielding cows. In the past a large number of bulls were required each year because of the small average size of the herds (Fig. 12.8) and the short time bulls were kept in them (Fig. 12.9). This position has changed. Herds are increasing in size (Fig. 12.8) and the use of artificial insemination, by which a bull can serve up to 80 000 cows a year instead of about 35 by natural mating, means that very few bulls are now required and better selection of bulls can be made. Grading up the commercial dairy cattle of the country by artificial insemination from proven bulls, as demonstrated in Fig. 12.10, is the first stage in the development of an improved national dairy herd. Selection schemes within breeds are the most important development now.

The bull is most important in breeding for milk (see Fig. 11.7). No matter whether the cows in the herd are good or bad to start with, the

Table 12.10   Cumulative feed energy cost of milk from birth to various ages of cow.

Age of cow (years)	Cumulative milk yield (kg)	Cumulative feed* energy eaten	Feed energy per kg of milk
$3\frac{1}{2}$	4 676	99 600	21.3
$5\frac{1}{2}$	16 366	204 500	12.5
$7\frac{1}{2}$	28 523	311 800	10.9
$9\frac{1}{2}$	40 680	419 200	10.3
$11\frac{1}{2}$	52 370	524 100	10.0

* Megajoules of metabolizable energy

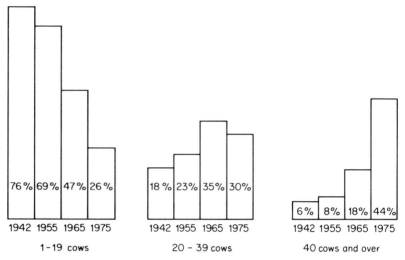

Fig. 12.8 Proportions of small, medium and large herds in 1942, 1955, 1965 and 1975. (Modified from *Dairy Herd* (*1965*) *Census*. Milk Marketing Board, 1966; and Milk Marketing Board, U.K. Dairy Facts and Figures, 1978.)

genetic composition of the herd after four generations will depend mainly on the genetic value of the four bulls used in succession, and but slightly on the genetic make-up of the original cows. Since a bull has to be 6 years old or over before he can become progeny tested, the number of naturally proven bulls in this country is not enough to supply the requirements of artificial insemination centres. In fact the artificial insemination system is the best means of proving a bull quickly, for few herds are large enough to do this. Selection schemes are run by artificial insemination organizations and by groups of private breeders.

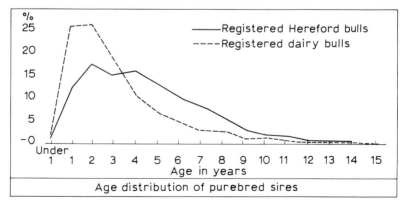

Fig. 12.9 Age distribution of bulls in beef and dairy herds. Most dairy bulls are disposed of before it is known (6 years old) how they are breeding for milk. (Buchanan Smith, A. D. (1931). *Journal of Agricultural Science*, **21**, 136.)

FOUNDATION COW

Mature weight, 372 kg

Milk, 2880 kg; Butterfat, 4.23%, 122 kg

50 PER CENT DAUGHTER

Mature weight, 471 kg; at birth, 32 kg

Milk, 5086 kg; Butterfat, 3.24%, 165 kg

75 PER CENT DAUGHTER

Mature weight, 464 kg; at birth, 29 kg

Milk, 7726 kg; Butterfat, 3.09%, 239 kg

87.5 PER CENT DAUGHTER

Mature weight, 553 kg; at birth, 42 kg

Milk, 7903 kg; Butterfat, 3.28%, 259 kg

Fig. 12.10 How milk production per cow can be improved by grading up successive generations to a good proven bull. (Woodward, T. E. and Graves, R. R. (1933). *Technical Bulletin of the United States Department of Agriculture,* No. 339.)

The main difficulty met with in the progeny testing of bulls is the time it takes before the bull's breeding value for milk production is known; the bull is usually about 6 years old before the first lactation records of his daughters are available. While waiting for the results of their progeny tests bulls must be laid off; the Milk Marketing Board has over 600 bulls laid off at any time. In countries where the population to be mated by artificial insemination is under 500 000 cows it is not usually necessary or profitable to keep bulls until their progeny test results are available. A more appropriate system then is to use the bulls to produce a limited number of offspring to provide a progeny test, then to take semen from the bulls into deep freeze for perhaps a year. Up to 80 000 doses of semen may be taken from a bull in a year and this would be sufficient for use in a small population of cows. At the end of semen collection the bulls can be slaughtered, so saving maintenance costs for five years. Semen taken from bulls which subsequently have a poor progeny test result can be destroyed.

Feeding and management as well as genetics have a large influence on a heifer's yield. In order to overcome the difficulty of variation in environmental conditions on different farms, the Milk Marketing Board have adopted a method of 'contemporary comparison' in which the first lactations of a bull's daughters are compared for the same years with those of daughters of other bulls in the same herd. In assessing the results, the actual yields of the daughters should also be taken into account, for a deceptively high index for a bull can be obtained if the other bulls with which he is compared are very poor ones. This is now done automatically in the calculation of the Improved Contemporary Comparison.

The contemporary comparison assesses a bull in terms of immediate practical interest—so much better or worse than the average bull. A good bull will be +250 kg or more; a bull poorer than average has a negative rating. The test replaces all other 'bull indexes' and makes possible the direct comparison of all bulls of the same breed in the country. It can be used for a series of small herds—provided that there is at least one heifer by some other bull to compare with the daughters of the test bull. It gives the same results in herds with high or with low yields—thus achieving the

Table 12.11   The progeny test of a single artificial insemination bull in herds at different levels of production. (From Robertson, A., Stewart, A. and Ashton, E. D. (1956). *Proceedings of the British Society for Animal Production*, p. 43.)

Range of herd averages (kg)	No. of daughters	Daughter average (kg)	Contemporary average (kg)	Difference, i.e. contemporary comparison
<3748	38	3814	3129	+684
3748–4216	53	4376	3715	+661
4217–4685	25	4742	4207	+534
>4685	58	5576	4816	+759

aim for which it was invented. This is illustrated by the figures in Table 12.11 which refer to an exceptionally good bull. It is clear that the daughter averages are conditioned entirely by the herd environment and give no indication of the merit of the bull. The difference between daughters and contemporaries, on the other hand, is independent of herd level. The final justification for the contemporary comparison method of progeny testing bulls is the relationship between tests on fathers and sons. Progeny-tested bulls, when mated to average cows, produce sons whose contemporary comparisons can be predicted (with at least 70 per cent of the expected accuracy) from those of their fathers (see Fig. 12.11).

The accuracy of a progeny test increases with the number of daughters, rapidly until the number reaches 20–25, after which additional daughters increase the accuracy more slowly. With the contemporary comparison the effective number of daughters is a weighted number rather less than the actual number of daughters. For instance, 30 actual daughters give a 'weight' of about 20 effective daughters, i.e. they give as much information as would 20 daughters all milking together in one herd and compared against a large number of daughters by other bulls. The Milk Marketing Board publishes the contemporary comparison of bulls who have a minimum of 20 effective daughters.

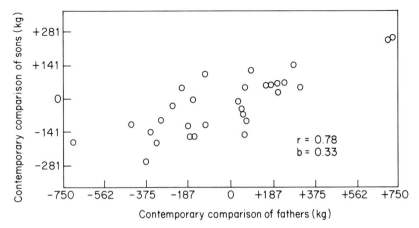

Fig. 12.11 Relationship between contemporary comparison of father and son, based on 30 A.I. Friesian bulls tested on at least 80 daughters and each having at least four accurately tested sons. Sons from fathers with progeny tests of more than +141 kg have, on average, contemporary comparisons better than zero. (Ødegård, A. K. and Robertson, A. (1967). *Acta Agriculturae Scandinavica*, **17**, 241.)

For butterfat percentage, daughter averages are satisfactory because, unlike milk production, this does not vary so much with feeding conditions. Naturally the yields of all and not some selected daughters of a bull should be used in calculating the progeny test.

The difficulty of environmental differences between farms can also be overcome by feeding 20 heifers, from each of several bulls to be tested, on standard rations on the same farm, as is done in Denmark, where the heifers are brought to the testing station a month or so before they are due to calve. This method was adopted by the British Oil and Cake Mills at Selby. It was found, however, that the heifers varied considerably in condition owing to different systems of rearing on the various farms from which they came. Because of this they tried bringing them in as calves, so that the heifers from different bulls would all be reared under similar conditions, a system which is used in Iceland. In countries with a large herd size, progeny testing stations are a much more expensive form of testing than the use of farm records. However, in countries with small herds progeny testing stations or special nucleus recording herds may be the only effective means of bull selection.

When good 'proven' bulls such as that shown in Fig. 12.12 are obtained, full use should be made of them for the improvement of dairy cattle, particularly by breeding further bulls for testing. These must be bred out of cows with a long lifetime record of high production. Because some high-yielding cows transmit their high yields better than others, an additional safeguard is to choose a cow which has produced high-yielding daughters. If bulls are selected in this way the next generation should produce animals not only with high yields but also with a long life and a good constitution. This is supported by New Zealand results (Fig. 12.13)

Fig. 12.12   A proven British Friesian bull, Hunday Falcon 5th. His contemporary comparison (milk bonus) is +234 kg. Yields of his daughters were:
  1027 first lactations averaged 4529 kg at 3.78% fat
  488 second lactations averaged 4604 kg at 3.77% fat
  427 third lactations averaged 5301 kg at 3.76% fat.
Daughters were graded 'very good' for temperament, ease of milking, dairy character, udder texture, teat shape and general conformation; 'good' for size, udder shape, teat position, and overall assessment; 'average' for feet and legs. (Courtesy of Milk Marketing Board.)

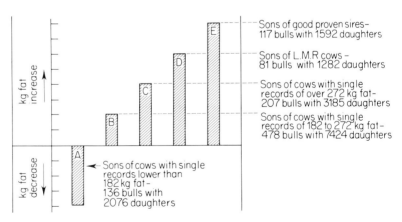

Fig. 12.13   Breeding results of bulls selected by different methods. From New Zealand statistics. It will be seen that if a proven bull cannot be obtained, the next best bull to use is the son of a proven bull out of a cow with a long lifetime production record (LMR cow). (Ward, A. H. (1945). *Sire Survey and Merit Register*. New Zealand Dairy Board, Wellington, 18.)

which show that the best method of selecting a young dairy bull is to take a son of a proven sire out of a cow with a long lifetime record of high production (Fig. 12.14), particularly if she has already produced good milking daughters. By this means one is breeding not only for production but also for constitution and livability, both essential factors in commercial milk production.

The chief criterion used in culling cows, and in progeny testing bulls, is milk yield—usually the 305-day first-lactation yield. The emphasis should be on the first lactation for the following reasons:

(1)   The first lactation gives a reasonably good indication of lifetime yield, i.e. repeatability of yield is fairly high.

(2)   Heritability of first lactation yield is higher than that of second

Fig. 12.14 Type of elite British Friesian cow considered suitable for mating to elite bulls to breed bulls for progeny testing. (Photograph by courtesy of K. C. Taylor, Preston.)
Terling Norah 255. Date of birth 22.12.1964.

Lact. No.	Age at Calving (yrs/mths)	Milk (kg)	Days	Fat %	Number of tests
1	2/8	5631	305	4.30	9
2	3/9	9056	305	4.16	10
3	4/11	9355	305	4.16	10
4	6/-	10064	305	3.97	9
5	7/10	12602	342	4.16	11
6	9/3	14058	377	4.49	11
7	10/6	13055	341	5.02	11
8	11/7	11576	315	4.05	10
9	12/7	14083	502	4.33	14
10	14/4	10884	257	5.57	8

lactation, i.e. the first lactation is less influenced by environmental factors (e.g. preceding dry period) and is thus a better indicator of a cow's genetic capacity.

(3)   In progeny testing, only for the first lactation is it possible to obtain a group of unselected daughters of a bull.

(4)   The results are available much sooner than if later lactations are awaited.

With some exceptions, which must be watched, bulls selected on the basis of first lactations are still the best bulls when judged on second and third lactations.

Owing to the negative correlation between milk and fat percentage, selection for milk yield will tend to lower fat percentage. This negative

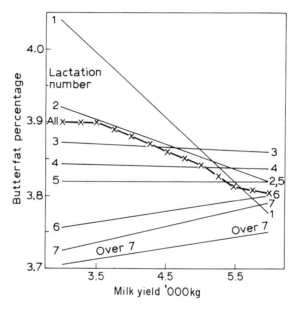

Fig. 12.15 The relationship between butterfat percentage and milk yield in different lactations. (Scottish Milk Marketing Board (1968). Milk Recording Services Report No. 3.)

correlation is less for second lactations and disappears in later ones (see Fig. 12.15). The trend can be kept in check by culling low-fat heifers and bulls while maintaining the main selection pressure on yield.

Figure 12.15 refers to individual cows within a breed; Table 12.12 shows that high-yielding breeds also tend to have lower butterfat and protein percentages.

The colour of the milk and butterfat has become important economically since milk has been sold in bottles. The colour is due to carotin, which is formed from the pigment associated with the green colouring matter of plants. In some breeds such as the South Devon, Guernsey and Jersey (Fig. 12.16) the colour of the body fat and butterfat is well developed, whereas

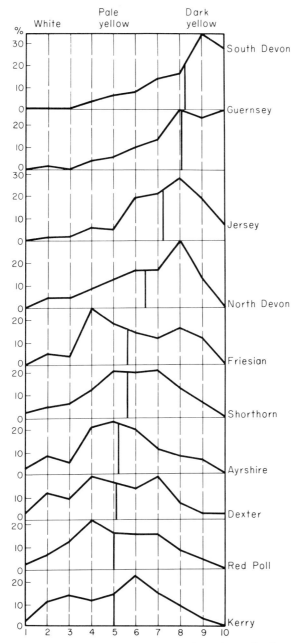

Fig. 12.16 Variability curves of the colour of the butterfat from individual cows of various breeds exhibited at the London Dairy Show. (Whetham, E. O. and Hammond, J. (1935). *Journal of Dairy Research*, **6**, 340.)

Table 12.12   Average yield of recorded herds in England and Wales, 1977–78, excluding first lactations. (*Milk Marketing Board, Report of the Breeding and Production Organization*, No. 29, 1978–79.)

	No. of herds	305-day milk yield (kg)	Butterfat percentage	Protein percentage
British Friesian	10 153	5586	3.74	3.24
Ayrshire	446	4950	3.89	3.34
Dairy Shorthorn	138	4851	3.59	3.28
Guernsey	399	3998	4.59	3.58
Jersey	449	3824	5.06	3.82

in many individuals of the other dairy and beef breeds the pigment is broken down in the liver and the fat is pale in colour.

In cattle this is probably a multiple-factor or blending character because all shades exist. In the absence of green food in the ration, all breeds produce butterfat dead-white in colour. As the amount of green food in the ration is increased, so the intensity of the colour increases to a maximum, which varies with the individual and breed, beyond which no amount of additional greenstuff will increase the colour. Thus, as for many other commercial characters, it is impossible to select and breed for butterfat colour unless optimum conditions are given for its formation.

In breeding tropical cattle for milk production 'temperament' is one of the factors to be considered. Milk is 'let down' by a reflex oxytocin discharge set up by suckling or stroking the teats (see p. 87). In the early days of domestication, and in unimproved herds today, this reflex will not take place except in the presence of the calf (Fig. 12.17). By selecting for a docile temperament, milk is let down without the presence of the calf. As has been shown by experiments in Jamaica and Trinidad, this character is inherited and can be transmitted through the bull (Hammond, 1932).

Fig. 12.17   The process of milking in the early days of domestication, showing how the calf was tied to the neck of the cow so that she would 'let down' her milk. Let-down is also being stimulated by the milker blowing into the vagina of the cow. Frieze from the temple at Tel-el-Obeid, Ur, about 3000 B.C. Frieze now in British Museum. (Duerst, J. U. (1931). *Grundlagen der Rinderzucht*. Berlin.)

Buffalo

In many parts of the world, but especially in Asia, the buffalo is an important animal for draught, for milk production and to a lesser extent for meat. There are about 140 million domesticated water buffaloes of two main types (Fig. 12.18) termed the Swamp and the River Buffalo, which will interbreed though they have different chromosome number (see Table

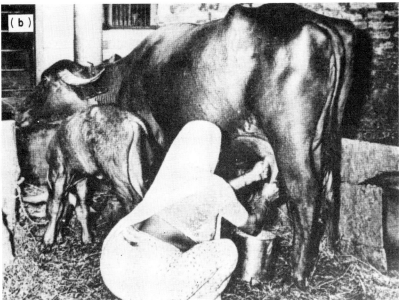

Fig. 12.18   Swamp and River Buffalo. (a) Feral Swamp buffalo undergoing domestication (Northern Territory, Australia), calf sucking from rear; (b) river buffalo (Surti breed, India), the calf is present to promote milk let-down. Improved feeding and management has increased yields by 30% on some Indian farms. (Food and Agricultural Organization (1977). *The Water Buffalo.* Rome.)

9.1). The former is predominantly a draught animal of the rice areas of the Far East whilst the latter is a dairy animal mainly found in India and Pakistan. There are many herds of buffalo ranging from 250 kg to 1000 kg mature weight. Milk production may vary from low yields of 1500 kg to high yields of over 5000 kg in lactations extending to 400 days. Buffalo milk contains more fat and protein than cows milk (Table 12.13). The buffalo female reaches puberty and sexual maturity later than cattle heifers of European or tropical breeds, and has a longer gestation period of between 300 and 330 days. The heritabilities of characters such as fertility, milk production and composition and growth rate in the buffalo are thought to be similar to those for the same characters in cattle. Buffaloes have less variation in body colour and there has been less interest in

Table 12.13 Composition of milk of domesticated farm animals. (From Kon, S. K. (1959). *Milk and Milk Products in Human Nutrition*. Food and Agricultural Organization, Rome.)

Type of milk	Fat percentage	Protein percentage	Lactose percentage	Total solids percentage
Buffalo	7.5	3.8	4.9	16.8
European cow	3.8	3.5	4.8	12.8
Zebu cow	5.0	3.2	4.6	13.5
Sheep	7.5	5.6	4.4	18.4
Goat	4.5	3.3	4.4	13.2

establishing breed societies and herdbooks than for cattle. Most buffalo are kept in small herds and artificial insemination is not widespread. For these and other reasons there have been few attempts to select the buffalo for important commercial characters. It is a reasonable assumption that selection methods which have been successful with cattle will be equally appropriate for buffalo. Thus improvement of fertility may be achieved by crossbreeding to benefit from hybrid vigour, and of growth rate can be based on performance testing and individual selection. Improvement of carcase quality would have to include some progeny testing. Milk yield and quality can be improved by progeny testing males on the first lactation records of their daughters. Small farm herd size, lack of AI and of recording may make this difficult except in large herds especially established for the purpose, as with the dairy cattle progeny testing stations in Denmark. In farm herds individual selection for milk yield has been tried but the rate of gain has been low and of the order of 0.1 per cent per year. Genetic gains of 0.5 to 1.0 per cent per year can be expected in schemes in large herds using progeny testing.

Beef cattle

Most of the beef in the world is produced from purebred and cross bred beef cattle. Beef cows are grazed under extensive, often hard and poor, range conditions and mated by natural service by bulls running with the cows on range. The calves produced are usually weaned at 6 months to a year and may be raised to slaughter either on improved pasture or on silage and concentrate feeds in a feedlot. Selection has been concerned mainly with improving the growth and carcase character of the bulls used, so that their offspring will perform better on pasture or on feedlot. It is difficult to select for improved beef cows because recording the performance of cows is not easy under range conditions. What little selection has been done has concentrated on selecting those cows which produce the greatest weight of weaned calf per year. Such a selection criterion is a composite of several characters including fertility of the cow, calf survival and growth, and cow milk yield to the calf.

Choice of breed allows selection of breed of cow, which can cope with the particular range conditions, and of breed of bull, which can survive to mate cows on range and produce fast growing, good carcase, calves for slaughter. Under many circumstances the choice of cows and bulls involves at least two breeds and sometimes three or more. Many of the cows are the cross of two breeds in order to obtain the benefits of hybrid vigour for reproductive characters and calf survival. A very large breed comparison is in progress at the U.S. Department of Agriculture Clay Center, Nebraska, in which the main beef purebreeds and crossbreeds are being assessed for the important beef characters.

In Europe beef production from beef-bred cattle (purebred or crossbred) is limited mainly to the marginal land, for the overhead charges on breeding cows on good land are too high under the one-calf one-cow system of rearing. Production of a twinning breed of beef cattle would, as in the case of sheep, reduce these overhead charges, but until this is done most of our beef must be supplied from other sources.

A large amount of beef in Europe, particularly in the U.K., comes from dairy cows which are either purebred (mainly Friesian) or crossbred (beef breed sire × Friesian cow). The calves used in this form of production are a by-product of the dairy industry. Though in the U.K. these calves have been mainly steers, the majority in mainland Europe are bulls because they produce leaner carcases to suit market needs and because they grow faster and convert feed to meat more efficiently. Improved methods of rearing will do much to reduce the costs of production and the quantity and quality of beef from dairy calves. By using first crosses, however, it is possible to utilize the poorer dairy cows to breed cattle suitable for beef. Not all the cows in a dairy herd are required to breed replacement heifers, and if in a herd of 100 cows 60 are inseminated by a good dairy bull the 40 worst can be inseminated by a good beef bull to supply calves for beef purposes. By such methods the dams of the replacements of the dairy herd are selected and the culls used for beef. For such a system of crossbreeding, the beef breeds most often used are those like the Hereford, Charolais and Aberdeen Angus, which have dominant characters and so colour-mark their calves (Fig. 12.19). The Charolais and the Simmental are particularly useful for increasing growth when crossed on the smaller dairy breeds (Fig. 12.20). In buying suckler weaned calves, a premium is paid for white-faced

Fig. 12.19   How beef bulls of breeds with dominant colours will colour-mark their calves and so distinguish them from dairy or dual-purpose animals. A Guernsey cow that was mated with an Aberdeen-Angus bull and with a Hereford bull, and produced twins, one by each bull. (Vieth, E. L. (1940). *Journal of Heredity*, **31**, 306.)

or yellow-dun coloured calves because it is clear that they are sired by a beef bull.

In these beef production systems it is possible to improve beef characters by selecting for better beef breed sires for crossbreeding. In addition it may be possible to select for beef as well as milk characters in dairy cows. However, the benefits of this are by no means proven. Faster growth rate in dairy cows may lead to larger cows with a higher feed maintenance but not necessarily with a proportionately higher milk yield. The improvements for beef may therefore be less than the losses in feed conversion for milk. Therefore beef improvement is best concentrated on selection of beef breed sires.

The most important characters to determine profitable production in beef cattle are growth rate, feed consumption and conversion, and carcase quality. The heritabilities of these characters (see Table 10.1) are medium to high and the same principles of selection hold as in breeding for lamb and pig meat. Growth rate of beef cattle can be measured in both cows and bulls. Beef cattle are often kept in small herds and their generation interval is already long. Performance and progeny test facilities are not interchangeable. Therefore growth rate can be most rapidly improved by mass selection (or individual selection, which is the same thing). In view of the small size of herds, breeders need to buy-in bulls to avoid inbreeding. To compare their own bulls with those from other herds their growth rate must be compared under the standard conditions of a performance-testing station.

Fig. 12.20   Beef potential of Ayrshire × Charolais. Two steers that were first and second prize winners at Gloucester Fatstock Show 1963. Details were:

	Age (days)	Liveweight (kg)	Deadweight (kg)	Killing out (%)
First steer	347	489	291	60.1
Second steer	345	458	263	57.5

(Milk Marketing Board. *Report of the Breeding and Production Organization*, No. 14, 1963–64.)

There is therefore no case for progeny testing 'breeders' bulls' for growth rate. There may be a case for progeny testing them for carcase characters, but only if the progeny test can be obtained from field records. It would not justify building special progeny-testing stations or detracting from performance-testing facilities.

The heritability of weight gain is fairly high (see Table 10.1). Therefore the bull's own performance gives a reasonably accurate estimate of his breeding value for growth. Rate of gain is correlated with efficiency of feed utilization so that, according to Pierce *et al.* (1954) for each increase of 0.1 kg gain per day over the average, 23 kg digestible nutrients for each 100 kg liveweight are saved. According to Swiger *et al.* (1965) selection for 'final weight alone is 90 per cent as efficient as a selection index including weaning weight, post weaning gain, feed consumption and fatness of carcase, but feed consumption and fatness would increase'.

Hence the importance of measuring feed consumption in the performance test. If carcase qualities are also important, a progeny test must be employed at present. However, research is in progress to find ways, such as ultrasonic techniques and body scanners, of measuring carcase quality in the live animal, and if successful this would largely do away with the need for progeny tests.

Performance testing of beef bulls can be carried out in three ways. The first, known as time constant testing, involves measuring the animals' growth rate over a certain time period, say of 150 or 200 days, irrespective of the age and weight of the individual animals at start of test. The second method, weight constant testing, is the measurement of the time it takes for animals to grow over a specified weight range, for example from 250 to 400 kg. The third method, age constant testing, is the weight gain of animals over a specified age range such as 150 to 400 days of age. For each method it is possible to measure feed intake. The second method has several advantages, particularly if the weight range for test is equivalent to the weight range for commercial beef animals. Poor performing animals can be culled at an early stage in the test if performance in early test is closely related to final test result. This means that more animals can be tested in the facilities available and the generation interval may also be reduced. In the United Kingdom bull performance tests are run by the Meat and Livestock Commission at 5 stations, accommodating in total about 400 bulls annually. Growth rate and feed intake of individual bulls is measured, using age constant testing for a period from about 150 to 400 days of age. The whole period of the test is considered as an equilibrating period during which the effects of different farm treatments before weaning and start of test can wear off. In the U.S.A., time constant testing is used in which bull calves are taken at weaning from their dams and, under uniform conditions of feeding, rate of gain and feed consumption is measured for 140, 196 and 240 days. At the end of test the bulls are classified according to conformation and finish. The full effect of performance testing on breed improvement can only be obtained when the best of the tested bulls are used to breed young bulls for testing in the next generation.

Tradition has favoured selection in a high nutrition environment, on the theory that this will show the extent to which more valuable parts of the carcase are capable of developing. Under range conditions where feeding is often poor after weaning, little progress is apparent from breeding, and such areas import bulls regularly to keep up the beef qualities of their stock. These bulls are produced in stud farms under conditions of unlimited nutrition, nurse cows frequently being used until they are about 18 months old so that proper selection can be made of bulls which show greatest development of beef qualities. One disadvantage of this system is that the bulls which grow best under conditions of intensive feeding are not necessarily the ones whose progeny grow best under more extensive conditions. In the U.K. it is probably satisfactory to performance test bulls on a high concentrate ration because their progeny, though on pasture, can also experience a high nutritional plane. But bulls selected in this way will probably not give the progeny best adapted to dry tropical ranges.

In the United Kingdom the MLC, in addition to performance testing stations, has provided a beef recording system used both by pedigree breeders and by commercial beef producers. This system has provided comparative breed information on growth rate. Comparisons of feed conversion for breeds and crossbreeds cannot usually be obtained from farm records, so the MLC have set up two Commercial Beef Evaluation Units for this purpose and have also obtained much information on breed carcase quality.

For a beef production system growth rate is closely related to profitability. The growth rates of different breeds of bull are a good indication of the comparative weight of their offspring from suckler cows and in the U.K. the ranking of breed performance is much the same irrespective of the farm type (see Table 12.14).

Table 12.14 400-day weight of beef bulls and the effect of sire breed on the 200-day weight of their calves. (From Baker, H. K. (1978). *Breeds and breed crosses for beef production.* George Scott Robertson Memorial Lecture, Queen's University of Belfast.)

400-day weight of pure bred bull (kg)	Sire breed	200-day calf weight (kg) Type of beef herd		
		Lowland	Upland	Hill
551	Charolais	240	227	205
532	Simmental	232	222	198
520	South Devon	231	221	200
460	Devon	225	215	191
510	Lincoln Red	222	214	189
445	Sussex	215	207	186
454	Limousin	215	204	186
424	Hereford	208	194	184
387	Aberdeen-Angus	194	182	176

The choice of beef bulls for use in artificial insemination presents different problems. At the moment they are used almost exclusively to produce commercial beef cattle—either crossbred breeding females or steers for slaughter. This means that criteria such as rate of genetic improvement are inapplicable. The maximum economic advantage will come from selecting accurately the bulls transmitting the economically important characters of growth rate, feed-utilization efficiency and carcase quality, and using them on a large scale.

The Milk Marketing Board in England and Wales has a special farm for progeny testing its beef bulls used for artificial insemination, and the Meat and Livestock Commission is using the records from its beef-recording service to progeny test other bulls used in artificial insemination.

The progeny test is easier to apply in beef than in dairy cattle, for beef qualities are visible in both sexes and at all ages, so that beef bulls can be proven at a much earlier age than dairy bulls. In progeny testing for beef, at least 20 steers from each bull are taken at weaning for beef breeds, and as calves for dairy breeds, and reared, preferably by individual feeding, under similar conditions; an appraisal is made of their carcases at slaughter. In these tests the question arises as to whether the steers should be killed at a certain degree of finish, at a certain age or at a certain weight. The first criterion is difficult to estimate; the second is easy to operate but may give some carcases which are too fat (or heavy) and some that are too lean (or light); the third has the advantage that carcase measurements can then be compared directly as in pig progeny testing, and is more comparable to commercial practice.

One of the chief difficulties in making progeny tests for beef is in evaluating the carcase qualities. The MLC have introduced a beef carcase classification scheme. The carcases are described by the following characteristics:

Weight—to include or exclude kidney, knob and channel fat

Sex—steer, heifer, cow or bull

Fatness—based on one of five fat classes according to the level of fat cover, with 1 indicating the leanest and 5 the fattest class (see Fig. 12.21). Class 3 is subdivided into 3L and 3H, with low and higher fat content.

Conformation—based on one of five classes with 5 indicating very good and 1 indicating poor conformation. About 1 per cent of carcases are very poor and classed as Z. The class is determined by a visual appraisal of shape, taking into account carcase thickness and blockiness and fullness of the round. Fatness is not taken into account. (See Fig. 12.22).

Classification is carried out by experienced staff to a national standard. About 21 per cent of carcases in 1979 were classified as average, 3L fatness 3 conformation. The market in the U.K. is tending to require a carcase with a classification of 2 fatness 4 conformation, with the European market requiring even leaner carcases and the U.S. market requiring fatter ones.

Poultry

In breeding for egg production the chief means of improvement is the progeny test. This is particularly so in the case of cocks, in which production cannot be measured directly. For example, it has been found better to breed from a hen which herself only produces say 150 eggs if she is the progeny of a cock whose daughters average 200 eggs, rather than from a 250-egg hen from a cock whose daughters only average 150 eggs. Usually the progeny test is modified into a sib test; that is, instead of waiting for the results of a progeny test a son of a progeny-tested cock is used, i.e. the brother (sib) of high-producing hens. Hens are selected according to their own yield and that of their sisters. Breeding from progeny-tested cocks and hens also ensures selection for constitution and livability, qualities which have been much neglected in the poultry industry in the past. Since such methods are costly and laborious, they are usually confined to large-scale breeders who supply breeding stock to those who are producing eggs for hatcheries, which in turn distribute day-old chicks to commercial producers.

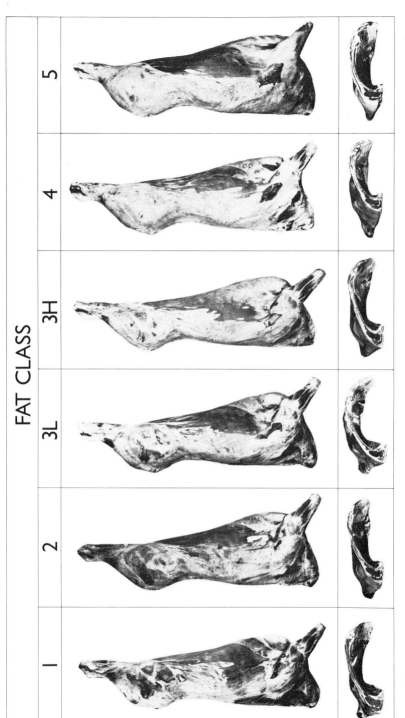

FAT CLASS

1	2	3L	3H	4	5

Fig. 12.21 The scale of fatness of beef carcases as used by the Meat and Livestock Commission, U.K. (MLC, Milton Keynes.)

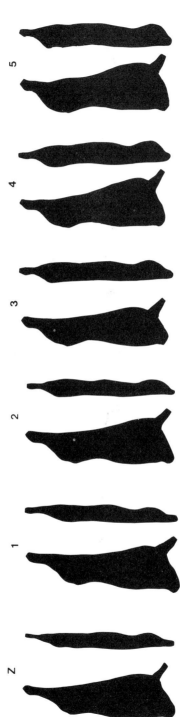

Fig. 12.22 The scale of conformation of beef carcases as used by the Meat and Livestock Commission, U.K. (MLC, Milton Keynes.)

Naturally many different qualities go to build up high egg production, a constitution suited to the environment not being the least important.

Egg production has increased under selection chiefly by reduction in the age at first egg (earlier sexual maturity) but also by increase in intensity of lay (reduction in broodiness and in laying pauses). Unfortunately, as egg yield increases egg size tends to decline, owing to the negative genetic correlation between these two traits. Egg weight must therefore be considered in selection as well as egg number. To increase efficiency of feed use, body weight must be held constant or reduced. Shell thickness is also important and, to a less extent, internal egg quality—firmness of white and absence of blood and meat spots.

Yolk colour, like butterfat colour previously mentioned, is an inherited character and can be bred for in the same way. In Europe dark yolks are preferred, in America light ones. Shell colour is also inherited and is probably controlled by several pairs of genes, some of them acting cumulatively to produce darker shades. In Britain, brown shells are preferred; in most of the U.S.A. and the Mediterranean area, white ones.

For commercial egg production crosses are now widely used, as the hybrid vigour of the first cross usually gives a high uniform level of egg production. For obtaining this hybrid vigour, however, much depends on crossing two or more strains which 'nick' well, that is, which contain characters complementary to one another. Such strains can be built up by progeny testing the cocks of one breed on the hens of the other breed, and then breeding such progeny-tested cocks back to hens of their own breed to produce a strain which will give good results when crossed with the other breed. This is known as recurrent selection.

In recent years the broiler industry has increased greatly. For this a bird of 1 to 2 kg at 7 to 9 weeks old is produced by feeding on a ration with a high energy content. This has entailed selection for rapidly-growing and plump-breasted strains. Broilers must also have white plumage and rapid feathering to avoid the dark marks of unerupted feathers under the skin. White skin and legs are preferred in Britain, yellow skin and legs in America. White legs are dominant to yellow legs; this is a simple segregating character and can easily be manipulated.

For both egg and broiler production, crossing is the usual breeding system. This may be a simple 2-way cross (A × B) but more usually it is necessary to employ a 3-way cross (A × BC) or a 4-way cross (AB × CD) to ensure that the dams (or both parents) are crossbreds and therefore the maximum benefit from hybrid vigour is gained in the multiplication of stocks for commercial sale.

General conclusions

Many of the unwanted characters (such as defects and deformities) and the fancy points (such as the hair colour, horns, etc.) are mutant characters and can be bred out, or bred for, quite easily by using Mendelian methods.

Most commercial characters, like milk yield, egg production and meat yield and conformation are, however, multiple-factor or blending characters and can best be bred for by selection on the basis of performance and progeny tests in an environment suited to develop the particular character in question; this is akin to the normal process of evolution.

Man can direct not only the evolution of his domestic animals, but also his own, by creating the environment—both mental and physical—in which he brings up the next generation.

References

HAMMOND, J. (1932). *Report on Cattle Breeding in Jamaica and Trinidad.* Publication No. 58, Empire Marketing Board, London.

LAND, R. B. (1974). Physiological studies and genetic selection for sheep fertility. *Animal Breeding Abstracts*, **42**, 155.

MORE O'FERRAL, G. J. and CUNNINGHAM, E. P. (1974). Heritability of racing performance in Thoroughbred horses. *Livestock Production Science*, **1**, 87.

New Zealand (1969). Research in the New Zealand Department of Agriculture 1967-68. *Annual Report of the Research Division (MMB), 1967-68.* Government Printer, Wellington.

PIERCE, C. D., AVERY, H. G., BURRIS, M. and BOGART, R. (1954). Rate and efficiency of gains in beef cattle. II. Some factors affecting performance testing. *Station Technical Bulletin, Oregon Agricultural Experiment Station*, No. 33.

SWIGER, L. A., GREGORY, K. E., SUMPTON, L. J., BREIDENSTEIN, B. C. and ARTHAUD, V. H. (1965). Selection indexes for efficiency of beef production. *Journal of Animal Science*, **24**, 418.

VARO, M. (1965). Some coefficients of heritability in horses. *Annales Agriculturae Fennicae*, **4**, 223.

WALLACE, L. R. (1964). The effect of selection for fertility on lamb and wool production. *Proceedings of the Ruakura Farmers Conference Week, 1964.*

Further reading

COCKRILL, W. ROSS (1974). *The Husbandry and Health of the Domestic Buffalo.* Food and Agriculture Organization of the United Nations, Rome.

EVANS, J. W., BORTON, A., HINTZ, H. F. and VAN VLECK, D. L. (1977). *The Horse.* W. H. Freeman & Co., San Francisco.

KEMPSTER, A. J., CUTHBERTSON, A. and HARRINGTON, G. (1982). *Carcase Evaluation in Livestock Production, Breeding and Marketing.* Granada. St Albans.

OWEN, J. B. (1976). *Sheep Production.* Bailliere Tindall, London.

POND, W. G. and MANER, J. H. (1974). *Swine Production in Temperate and Tropical Environments.* W. H. Freeman and Co., San Francisco.

PRESTON, T. R. and WILLIS, M. B. (1979). *Intensive Beef Production*, 2nd edition. Pergamon Press, Oxford.

SCHMIDT, G. H. and VAN VLECK, L. D. (1974). *Principles of Dairy Science.* W. H. Freeman and Co., San Francisco.

United States Department of Agriculture (1977). Beef Cattle Breeding. *Agricultural Information Bulletin (U.S.D.A.)*, No. 286, 76 pp.

Author Index

Subject Index